Advances in Intelligent Systems and Computing

Volume 1321

The series "Advances in Intelligent Systems and Computing" contains publications on theory, applications, and design methods of Intelligent Systems and Intelligent Computing. Virtually all disciplines such as engineering, natural sciences, computer and information science, ICT, economics, business, e-commerce, environment, healthcare, life science are covered. The list of topics spans all the areas of modern intelligent systems and computing such as: computational intelligence, soft computing including neural networks, fuzzy systems, evolutionary computing and the fusion of these paradigms, social intelligence, ambient intelligence, computational neuroscience, artificial life, virtual worlds and society, cognitive science and systems, Perception and Vision, DNA and immune based systems, self-organizing and adaptive systems, e-Learning and teaching, human-centered and human-centric computing, recommender systems, intelligent control, robotics and mechatronics including human-machine teaming, knowledge-based paradigms, learning paradigms, machine ethics, intelligent data analysis, knowledge management, intelligent agents, intelligent decision making and support, intelligent network security, trust management, interactive entertainment, Web intelligence and multimedia.

The publications within "Advances in Intelligent Systems and Computing" are primarily proceedings of important conferences, symposia and congresses. They cover significant recent developments in the field, both of a foundational and applicable character. An important characteristic feature of the series is the short publication time and world-wide distribution. This permits a rapid and broad dissemination of research results.

Indexed by SCOPUS, DBLP, EI Compendex, INSPEC, WTI Frankfurt eG, zbMATH, Japanese Science and Technology Agency (JST), SCImago.

All books published in the series are submitted for consideration in Web of Science.

More information about this series at http://www.springer.com/series/11156

Wida Susanty Haji Suhaili ·
Nor Zainah Siau · Saiful Omar ·
Somnuk Phon-Amuaisuk
Editors

Computational Intelligence in Information Systems

Proceedings of the Computational Intelligence in Information Systems Conference (CIIS 2020)

Springer

Editors
Wida Susanty Haji Suhaili
School of Computing and Informatics
Universti Teknologi Brunei
Gadong, Brunei Darussalam

Nor Zainah Siau
School of Computing and Informatics
Universiti Teknologi Brunei
Gadong, Brunei Darussalam

Saiful Omar
School of Computing and Informatics
Universiti Teknologi Brunei
Gadong, Brunei Darussalam

Somnuk Phon-Amuaisuk
School of Computing and Informatics
Universiti Teknologi Brunei
Gadong, Brunei Darussalam

ISSN 2194-5357 ISSN 2194-5365 (electronic)
Advances in Intelligent Systems and Computing
ISBN 978-3-030-68132-6 ISBN 978-3-030-68133-3 (eBook)
https://doi.org/10.1007/978-3-030-68133-3

This Springer imprint is published by the registered company Springer Nature Switzerland AG
The registered company address is: Gewerbestrasse 11, 6330 Cham, Switzerland

Preface

This book contains the proceedings of the 4th Computational Intelligence in Information Systems (CIIS 2020) Conference held during 25–27 January 2021.

Due to the COVID-19 pandemic, this conference which was initially planned in 2020 has to run in 2021 to accommodate all the uncertainties, and with the travel restriction for our international participants, we have to opt to conduct the conference virtually. Despite this situation, we are pleased to have received 64 submitted articles from 18 countries including countries such as Canada, Bulgaria, Brazil, Uganda and a bigger number from Asia including ASEAN Member States, Japan, Taiwan, Pakistan, Sri Lanka, India, Bangladesh and from the middle east, from Saudi Arabia and Iraq and finally from Uganda. Fifty-five go through the review process while nine have to be omitted due to the negligence of the submission which did not fulfil the requirements of the conference paper requirements. These papers have gone through a double-blind review with reviewers coming from 12 countries with some from the USA, South Korea and Australia. Out of these 55 submissions, 23 paper submission (49%) have been selected to be included in this book.

This year's theme highlights the digital transformation towards Sustainable Development Goals where we have papers on digital transformation, computational intelligence in application, networking domain and paper on COVID-19. We are also pleased that we have a number of young researchers that have chosen our platform to elevate their publication profile. We would like to thank all authors for submitting their papers and finally completing the process for publication and gratefully acknowledging our reviewers for their time in reviewing these submissions on time. We will look forward for more collaboration with other research communities in the next event.

We would like to express our sincere gratitude to AUN/SEED-NET JICA for their involvement and contribution in introducing us to both of our distinguished keynote speakers, Prof. Kenji Araki and Prof. Kazuhiko Hamamoto, from Japan and to AITI for connecting us to two other distinguished speakers: Dr. Afif Osseiran from Ericsson and Dr. Ksiuk Kweon from Samsung Electronics. Most importantly, we would like to convey our deepest appreciation to all members of the working committee for their relentless effort in ensuring the success of this conference despite the current pandemic situation that affects most of the planning that requires us to work outside our norms. We thank the publisher for publishing the proceedings, members of the steering committee for their valuable advice and last but not least, Hjh Zohrah binti Haji Sulaiman, our Vice Chancellor.

We would also like to acknowledge our respective sponsors for their assistance indirectly making CIIS 2020 a smooth one despite the pandemic.

Wida Susanty Haji Suhaili
Nor Zainah Siau
Saiful Omar
Somnuk Phon-Amuaisuk

CIIS 2020 Conference Committee

Honorary Chair/Advisor

Hajah Zohrah binti Haji Sulaiman	Vice Chancellor, UTB

Steering Committee

Chairperson

Mohamed Hasnain Isa	Assistant Vice Chancellor Research, UTB

Members

Hajah Noor Maya binti Haji Md. Salleh	Assistant Vice Chancellor Academic, UTB
Haji Ady Syarmin bin Hj Md Taib	Assistant Vice Chancellor Industry & Services, UTB
Lim Chui Hua	Registrar and Secretary, UTB
Hamdani bin Haji Ibrahim	Acting Bursar, UTB
Au Thien Wan	Dean Graduate Studies and Research, UTB
Mohamad Saiful bin Haji Omar	Dean School of Computing and Informatics, UTB
Hajah Nor Zainah binti Haji Siau	Deputy Dean School of Computing and Informatics, UTB
Wida Susanty binti Haji Suhaili	Thrust leader for Digital and Creativity, UTB

International Advisory Board

Eran Edirisinghe	Loughborough University, UK
Hussein A. Abbass	University of New South Wales, Australia
Irwin King	Chinese University of Hong Kong, China
Ah Hwee Tan	Singapore Management University, Singapore
Laszlo T. Koczy	Budapest University of Technology and Economics, Hungary
Ali Selamat	Universiti Teknologi Malaysia, Malaysia
Chris Phillips	Newcastle University, UK
Francis Chin	University of Hong Kong, Hong Kong
Jeff Haywood	University of Edinburgh, Glasgow, UK
Ian Ruthven	Strathclyde University, Glasgow, UK
Peter Haddawy	Mahidol University, Thailand
Jun Kyun Choi	Korea Advanced Institute of Science and Technology, Korea
Robert H. Deng	Singapore Management University, Singapore
Aduwati Sali	Universiti Putra Malaysia, Malaysia

Working Program Committee

Chairman and Co-chair

Wida Susanty binti Haji Suhaili
Hajah Nor Zainah binti Haji Siau

Program Chairs

Somnuk Phon-Amnuaisuk
Mohamad Saiful bin Haji Omar

Secretariat

Jennifer Voon Hyuk Hiong (Head)
Haji Irwan Mashadi bin Haji Mashud
S. H. Shah Newaz
Ravi Kumar Patchmuthu
Ibrahim Venkat
Noreffa Nabilla binti Abd Aziz
Ak Mohammad Salihin Pg. Haji Abdul Rahim

Technical Committee

Somnuk Phon-Amnuaisuk (Head)
Au Thien Wan
Haji Sharul Tazrajiman bin Haji Tajuddin
Ibrahim Venkat

Ravi Kumar Patchmuthu
S. H. Shah Newaz
Hajah Nor Zainah binti Haji Siau
Wida Susanti binti Haji Suhaili
Mohamad Saiful bin Haji Omar
Asem Kasem

Poster Competition

Serina binti Haji Mohd Ali

Finance

Haji Sharul Tazrajiman bin Haji Tajuddin (Head)
UTB Finance Officer

Logistics

Haji Irwan Mashadi Haji Mashud (Head)
Estate's Officer

ICT Logistic

ICTC's officer

Ceremony and Digital Content Coordinator

Haji Idham Maswadi bin Haji Mashud (Head)
S. H. Shah Newaz
Nurulhidayati binti Haji Mohd Sani
Attachment Students

Multimedia

Mohd Effendy bin Hj Mohd Jeffery (Head)
Hanif Jumaat

Technical Moderators

Haji Sharul Tazrajiman bin Haji Tajuddin (Head)
Ahmad Ammar Ghaibeh
Ibrahim bin Edris
Noor Deenina binti Haji Mohd Salleh
Rahizah Binti Haji Abdul Rahman
Abdul Aziz bin Metassan

Web Masters

Rafidah binti Haji Tengah (Head)
Shiqah Natasha binti Hj Mohammad Hadi

Sponsorship and Promotion

Yeo Sy Mey (Co-head)
Au Thein Wan (Co-head)
UTB Finance Officers

Publicity

Hajah Siti Asmahlati Bolkini Binti A. Hj Ahmad (IPRO)

Publishing

Noor Deenina binti Haji Mohd Salleh

Invitation and Protocol

Haji Rudy Erwan bin Haji Ramlie (Head)
Registrar's Office

Souvenir and Certificate

Siti Rabiqah binti Ramli

Refreshment

Norhuraizah binti Haji Md Jaafar

Workshop facilitator

Asem Kasem (Head)
Ravi Kumar Patchmuthu
Chongomweru Halimu
Muhammad Tariq Siddique

Technical Program Committee

Abdelrahman Elfaki	University of Tabuk, Saudi Arabia
Abdollah Dehzangi	Rutgers University, USA
Adham Atyabi	University of Colorado Colorado Springs and Seattle Children's Research Institute, USA
Ammar Ghaibeh	Universiti Teknologi Brunei
Annapurani K.	SRM Institute of Science and Technology, India
Asem Kasem	Universiti Teknologi Brunei
Ashraf Khalil	Universiti Teknologi Brunei
Au Thien Wan	Universiti Teknologi Brunei
Azizi Ab Aziz	Universiti Utara Malaysia, Malaysia
Bok-Min Goi	Universiti Tunku Abdul Rahman, Malaysia
Chattrakul Sombattheera	Mahasarakham University, Thailand
Choo Yee Ting	Multimedia University, Malaysia
Daphne Lai	Universiti Brunei Darussalam
Derwin Suhartono	Bina Nusantara University, Indonesia
Fahmi Ibrahim	Universiti Teknologi Brunei

Hui-Ngo Goh	Multimedia University, Malaysia
Hyeontaek Oh	Korea Advanced Institute of Science and Technology, Korea
Ibrahim Venkat	Universiti Teknologi Brunei
Kah Ong Michael Goh	Multimedia University, Malaysia
Kannan Arputharaj	Vellore Institute of Technology, India
Keng Hoong Ng	Multimedia University, Malaysia
Kevin Wong	Murdoch University, Australia
Kok Chin Khor	Universiti Tunku Abdul Rahman, Malaysia
Lai-Kuan Wong	Multimedia University, Malaysia
Lim Tiong Hoo	Universiti Teknologi Brunei
Laszlo T. Koczy	Budapest University of Technology and Economics, Hungary
Lee Kian Foo	Multimedia University, Malaysia
M. Rakib Uddin	Universiti Teknologi Brunei
Mohamed Saleem Haja Nazmudeen	Universiti Teknologi Brunei
Mohammad Saiful Hj Omar	Universiti Teknologi Brunei
Myint Myint Sein	University of Computer Studies, Yangon
Nor Zainah Siau	Universiti Teknologi Brunei
Nyuk Hiong Voon	Universiti Teknologi Brunei
Peter Haddawy	Mahidol University, Thailand
Phooi Yee Lau	Universiti Tunku Abdul Rahman, Malaysia
Ravi Kumar Patchmuthu	Universiti Teknologi Brunei
Rehan Akbar	Universiti Tunku Abdul Rahman, Malaysia
Rusyaizila Ramli	Management and Science University, Malaysia
S. H. Shah Newaz	Universiti Teknologi Brunei
Serina Mohd Ali	Universiti Teknologi Brunei
Shah Rollah Abdul Wahab	Universiti Teknologi Malaysia, Malaysia
Shahid Anjum	Universiti Teknologi Brunei
Sharul Tajuddin	Universiti Teknologi Brunei
Somnuk Phon-Amnuaisuk	Universiti Teknologi Brunei
Sook Ling Chua	Multimedia University, Malaysia
Soon Jiann Tan	Universiti Teknologi Brunei
Connie Tee	Multimedia University, Malaysia
Tho Quan	Ho Chi Minh City University of Technology, Vietnam
Wan Zuha Wan Hasan	Universiti Putra Malaysia, Malaysia
Whee Yen Wong	Universiti Tunku Abdul Rahman, Malaysia
Wida Susanty Haji Suhaili	Universiti Teknologi Brunei
Yunli Lee	Sunway University, Malaysia

Contents

Digital Innovation Towards Sustainable Development Goals

Innovation Development and R&D Project Management in Science Organizations and Universities - Data-Driven Model and Analysis

Zornitsa Yordanova(✉)

University of National and World Economy, 8mi dekemvri, Sofia, Bulgaria
zornitsayordanova@unwe.bg

Abstract. This study examines the scope and manifestation of innovation development and R&D projects management in science organizations and universities from project and innovation management perspective. For the purposes of the research, a data-driven model is developed in Qlik application by applying datasets from Scopus database (an Advanced Analytical Literature Review-AALR is used). The data-driven model is based on the principles of systematic literature review method and steps on data from more than 360 science articles, included in Scopus database.

Based on the results, provided by the data-driven model and application, analyses conclude the scope of the researched topic and some knowledge gaps. The findings and contributions of the research are: 1) New insights to the topic of innovation development and R&D projects management in science organizations and universities from project and innovation management perspective and 2) Data-driven model and application for systematic literature review (based on word and context analysis).

Keywords: Innovation development · Project management · Science management · University management · Data-driven · Innovation projects

1 Introduction

The current research is motivated by the increasing trend of unsuccessful science projects related to innovation development [1]; [2], ready for commercialization and an extending gap between science and market needs [3]. Previous researches have already revealed insufficient project management skills in science organizations [4], weak knowledge about the full cycle of innovation development [5], absence of practical approach for developing innovations which solve real problems [6] and lack of knowledge for developing innovations with fast product-to-market innovation management approaches [7].

W. S. H. Suhaili et al. (Eds.): CIIS 2020, AISC 1321, pp. 3–12, 2021.
https://doi.org/10.1007/978-3-030-68133-3_1

The current paper aims at analyzing through a systematic literature review and application of a specially developed advanced analytics tool for word and context analysis the scope of innovation development and R&D projects management in science organizations and universities. The tool is Advanced Analytical Literature Review (AALR) built on Qlik software and performs a word and context analysis on words and combination of words within science papers title, abstracts and author key words. The systematic literature analysis covers 360 science papers based on different advanced searches in the Scopus database as: "project management", "science projects", "innovation management", "science", "university", "science organization", "R&D projects", "innovation", and "innovation projects".

The structure of the paper starts from reviewing the theoretical background of the innovation development and R&D in science organizations and universities and project management in this kind of organizations. Then a research design is presented and the results from performing the systematic literature review for stating the art of innovation development and R&D projects management in science organizations and universities. The relevance of the research is associated with an increasing need of new innovation developments based on scientific research results as well as the decreasing lifetime of innovation products [8] which additionally extend the need for more effective innovation development and management in science organizations and universities.

2 Theoretical Background

2.1 Innovation Development and R&D in Science Organizations and Universities

Rogers [9] defines the innovation-development process as all the decisions, activities, and their impacts that occur from recognition of a need or problem, through research, development, and commercialization of an innovation through diffusion and adoption of the innovation by users, to its consequences. Back in the 80s of the 20th century, still innovation development has not been largely analyzed and researched [10]. Since then, innovation development is amongst the hottest and most research innovation-related topics.

Innovation development in science organizations and universities has extended scope. The most recent research from 2019 and 2020 focus on the development of innovation capabilities, challenges and future research trend in the UAE high education [11], science and education as reliable segments for the diffusion of innovations and the functioning of technology parks as the most effective organizational and economic form of integration of science and production among all other innovative structures in the world (case of Ukraine) [12], university–industrial relations as the main factor in the development of education (case of polytechnic education) [13], the role of scientific and industrial cooperation in assessing the innovative potential of an industrial enterprise and the approach to evaluation through joint patent and licensing activities [14], etc. This paper proves the need to include an assessment of the level of scientific and industrial cooperation in the analysis of the industrial enterprise innovative potential. Innovation development in universities is also seen as a mean for leveraging social capital in university-industry knowledge transfer strategies [15] by understanding how social capital in university-industry partnerships affect knowledge transfer strategies, which impacts on collaborative innovation developments. In other research sources, the link between innovation development and science organizations and universities is generally limited to the establishment of an innovation ecosystem [16] and regions development. Especially for the role of universities in the regional development, this role has been seen in the emergence of a third role of universities that has re-shaped and transformed their two traditional functions of teaching and research [17].

Research and Development (R&D) is a common activity in science organizations and in some universities as well. A review of the latest literature reveals some case studies as those of the University of Oulu, Finland which purpose was to examine knowledge-creating interaction in developing an innovation in a multidisciplinary research community with hermeneutic phenomenology, to understand how previous experiences and future prospects shape the process and to examine the circumstances; which support or limit knowledge creation [18]. Most of the R&D research is discussing concrete science activities outside of the management. However, there are few science articles in the literature analyzing the management side of these processes and they are generally related to project management.

2.2 Projects Management in Science Organizations and Universities

Organizations are becoming more project-oriented [19], and projects are the preferred management instrument especially for implementation of new activities [20]. These statements also apply to research organizations and universities. Science and educational institutions usually use a traditional approach for project management (PM). The main knowledge sources of these kind of project management methodologies are the International Project Management Association (IPMA), the Project Management Institute (PMI) which publishes a book on the areas of knowledge in project management (A Guide to the Project Management Body of Knowledge, PMBOK Guide), European Project Management Organizations (EPMO), Association of project management (APM), etc. Most recognized methodology for project management in universities is reported to be PMBOK by PMI [21]. In the literature are also known some attempts for developing and evaluating a project management methodology for university-industry collaborative projects [22] but still they are based on traditional approaches. However, since university–industry projects face specific challenges in understanding and expressing the values required of project management in delivering stakeholder benefits, a framework for understanding, identifying and managing the values of project management in major university–industry R&D projects has been developed by Fernandes et al. [23]. The value of this framework is the provided stakeholders' perspective by identifying the respective PM values for each of six stakeholders' types/collaborations: university–industry consortium; university; industry; R&D external entities; funding entity and society. In the latest years, extensive processes for adapting some flexible project management methodologies and methods have also happening. Consecutively, an Agile project management in university-industry collaboration projects is more and more of the interests of researchers and some case studies are already presented in the literature [24].

Management of performance, leadership and communication are also part of some research papers recently as part of the project management in science organizations and universities. The topic of a professional project manager leading R&D projects instead of scientists is also part of the discussions. Still, most of these projects are managed by principal researchers which hypothesis is analyzed in a research by Cassanelli, Fernandez-Sanchez and Guiridlian emphasizing that the coordinator or project manager is carrying out the search for financing and to manage contracts, resources, cost, time, scope, risk and uncertainty, communication, stakeholders and so on, in addition to internal research activities [25].

3 Research Design

The research design of the paper for achieving data-driven analysis on the topic of innovation development and R&D projects management in science organizations and universities is based on a systematic literature review and the use of a specially developed model and analytical tool for data analytics.

3.1 Systematic Literature Review

For the purpose of defining comparatively wide and still focused scope of the research, an advanced search in Scopus database was first undertook. It aimed at scoping science papers which are relevant to innovation development, R&D and project management in science organizations and universities. The dataset of science papers received by these searches has been cleaned (by removing duplicates) and uploaded in the Advances Analytical Literature Review software by designing model for their common and generic analysis.

3.2 Model and Tool for Advanced Analytical Literature Review (AALR)

A special model and a tool have been designed and developed to serve the research. The tool is based on Qlik Sense application, which is amongst the best recognized business intelligence tools for data analytics. It is basically used for enterprise data analysis and its application in the current research through AALR is indeed an innovative method for deeper word and contextual analyzing of the topic. It aims at revealing hidden connections between science papers with different focus and different source which are usually misunderstood by the traditional approaches of literature analysis and review. AALR is very useful and extremely appropriate for interdisciplinary research in which the analysis includes literature from diverse science fields.

The tool AALR integrates all the results from the Scopus searches showed in the first step of the research design. AALR is configured with various filters by which the researcher can search and combine different words in order to compare and collate the use of different words within large amount of research papers in their titles, author keywords and abstracts. By doing this, the researcher is able to reveal hidden context, to discuss and analyze the use of words in their context, to make comparison and to identify links between different research. It might be also useful for defining knowledge

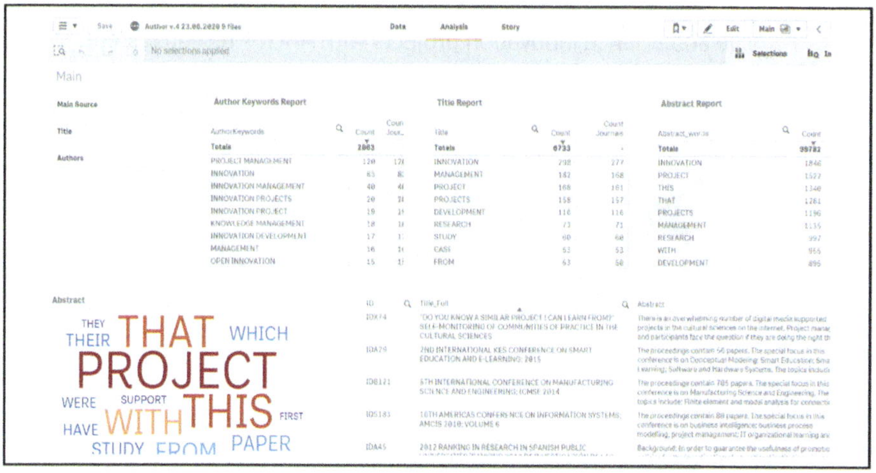

Fig. 1. Data mining tool AALR for word and context analysis of science papers

gaps. In the case of this research and the loaded data in the tool AALR, which is Scopus sourced, the tool has the look presented on Fig. 1.

The tool presents titles, author keywords and abstracts and it decomposes all the words of these components. There are search functionality on each of the sections for titles, author keywords and the abstracts which proposes an advanced search functionality to search by many words or words' fractions. These three filters are combining and the researcher can use three of them independently which will make him/her able to filter science papers on different filtering based on titles, author keywords and the abstracts. In addition, there is an option for the user to select different files from the search and some more features.

3.3 Analyses Performed

First, prior to using the specially developed tool AALR with the uploaded data from Scopus, a simple excel search was performed. An excel filter was added and a search function was started on the base of the titles of the data set. The search was based on science papers, containing the key words, subject of research in this paper. Several more combined searches were undertaken to reveal the work already done on the scope of innovation development and R&D projects management in science organizations and universities.

In the tool AALR, several analyses have been performed, based on word and context analyses. The consequence of the analyses is mainly described on Fig. 2.

The analyses performed with the tool AALR are present in the results' section.

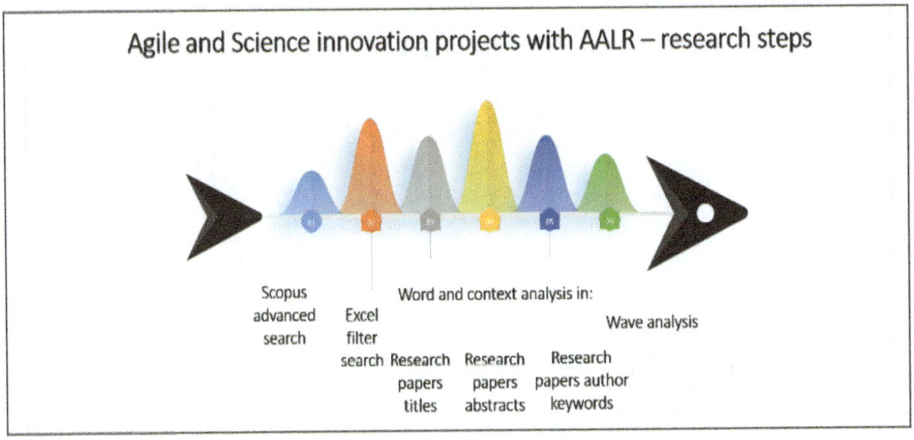

Fig. 2. Research design for a systematic literature review with AALR

4 Results and Discussion

The findings of the research bring insights to audience, interested in innovation development and R&D projects management in science organizations and universities from project and innovation management perspective. The data used is considered reliable since the model and the analyses are based on research papers from Scopus database. First natural exclusion criteria of the data analysis are by excluding any input from the advanced search performed in the Scopus database with these inclusion criteria in it:

TITLE-ABS-KEY ("innovation development") and TITLE-ABS-KEY ("science projects")

TITLE-ABS-KEY (*"innovation development"*) AND TITLE-ABS-KEY (*"science organization"*)

It turned out that no one research paper has been published covering the requirements of these two formulas. That means that there is no one research paper, containing simultaneously 'innovation development' and 'science projects' in title, abstract or authors' key words. This is also valid for papers, containing 'innovation development' and 'science organization' together. The research papers meeting the formulas are 839 in total, but because of the close topics within the nine different searches, there are duplicates. After a duplicates remove, the 'golden source' for the analyses contains 762 single research papers.

For making the scope clearer and focused, a filtering option is used to funnel the papers which have 'project management' and 'innovation' in their author keywords. The results by using the model and application AALR show these two terms have been used in 203 papers within the golden source. The results show that the most used words in abstracts of papers, dealing with 'project management' and 'innovation' are: development, knowledge, process, science, approach, engineering, model, success, factors.

The papers, referring to 'university' are 35 in total, containing 55 times the word 'university' in their abstracts (as it is presented in the wave analysis above). These 35 research papers are manually analyzed to summarize the current research activities. Only 3 of these focus on university innovation development presented on Table 1.

Table 1. Innovation in university

Authors	Title	Journal and affiliation	Year of publication
Hansen, Irina-Emily; Mork, Ola Jon; Welo, Torgeir [26]	Exploring framework for university-industry innovation projects: building collaborative knowledge platform	European conference on knowledge management Norwegian university of science and technology	2019
Tan Owee Kowang, Choi Sang Long & Amran Rasli [27]	Innovation management and performance framework for research university in Malaysia	International education studies Universiti Teknologi Malaysia	2015
Indrit Troshani; Giselle Rampersad; Carolin Plewa [28]	Organisational adoption of e-business: the case of an innovation management tool at a university and technology transfer office	International journal of networking and virtual organisations University of Adelaide	2011

By analyzing these three papers, it is turned out that innovation development practices in universities are discussed in the literature only as case studies or developing frameworks.

A next analysis is performed to reveal some project management flexible approaches used in universities for project and innovation management purposes. Within the AALR tool, containing the golden source of 762 papers, word filtering in the abstracts was performed (with the assumption that the abstracts hold the biggest quantity of information and they would indicate the usage of flexible project and innovation methodologies even though their usage is not the main focus of these researches). The used flexible project and innovation methodologies for revealing papers dealing with such are: 'agile', 'scrum', 'lean', 'six sigma'. In addition, in the author key words are selected 'innovation' and 'project management' and 'innovation' in the title so as to funnel the results at their maximum level to the most relevant ones. The results reveal three papers again which meet the set criteria. These are presented on Table 2.

The years of publication of the most focused research papers show increase in the interest towards innovation and project management in universities and science organizations. Meanwhile, the results also show a literature gap on innovation development in universities and science organizations as these papers are so few in quantity.

Table 2. Flexible project management for innovation in universities and science organizations

Authors	Title	Publishing source	Year of publication
Gregory C. McLaughlin, William R. Kennedy [29]	Innovation project management handbook	New York: productivity press	2016
Yordanova Z., Stoimenov N., Boyanova O., Ivanchev I. [30]	The long way from science to innovation – a research approach for creating an innovation project methodology	Lecture notes in business information processing, springer	2019
Philipp A. Lill, Andreas Wald and Ronald Gleich [31]	Agility and the role of project-internal control systems for innovation project performance	International journal of innovation management	2020

5 Conclusion

In conclusion of the presented analyses by using data model and specially developed advanced analytics tool, few findings might be highlighted:

- Project management in universities and science organizations is analyzed in the literature separately from innovation management and development
- Still very few applications of flexible project management models and approaches are analyzed in the science literature
- There is a knowledge gap in the literature on combining innovation management and project management for the specifics of innovation projects in science organizations and universities.
- Some commonly research topics along with innovation and project management in the context of science organizations and universities are: processes, models, performance, technology, collaboration, data and information.

The future work of the author in the context of this research will be focused developing a special framework/methodology for project management of innovations in science organizations and universities with all specifics of this proper environment and affecting factors. The framework will transfer knowledge from the flexible management approaches as Agile, Scrum, Six sigma, Lean Startup, User-centricity and Design Thinking, which are successfully applying in business projects especially for innovation development and innovation management.

Acknowledgments. The paper is supported by the BG NSF Grant No KP-06 OPR01/3-2018.

References

1. Etzkowitz, H., Zhou, Ch.: The Triple Helix: University–Industry–Government Innovation and Entrepreneurship, 2nd edn (2017). ISBN 9781315620183
2. Leahey, E.: From sole investigator to team scientist: trends in the practice and study of research collaboration. Ann. Rev. Sociol. **42**(1), 81–100 (2016)
3. Turel, O., Kapoor, B.: A business analytics maturity perspective on the gap between business schools and presumed industry needs. Commun. Assoc. Inf. Syst. **39**, 6 (2016). http://aisel.aisnet.org/cais/vol39/iss1/6
4. Pinto, J., Winch, G.: The unsettling of "settled science:" the past and future of the management of projects. Int. J. Project Manage. **34**(2), 237–245 (2016)
5. Hauge, S., Pinheiro, M., Zyzak, B.: Knowledge bases and regional development: collaborations between higher education and cultural creative industries. Int. J. Cult. Policy **24**(4), 485–503 (2018). https://doi.org/10.1080/10286632.2016.1218858
6. Pearce, J.: Teaching Science by Encouraging Innovation in Appropriate Technologies for Sustainable Development (2019)
7. Stock, G., Greis, N., Fischer, L.: Organisational slack and new product time to market performance. Int. J. Innov. Manage. **22**(04), 1850034 (2018)
8. Mylnikov, L., Fayzrakhmanov, R., Kurushin, D.: Information support of project management processes in the network of research organizations and technological companies. Astra Salvensis **6**, 603–612 (2018)
9. Rogers, E.: Diffusion of Innovations, 5th edn. Free Press, New York
10. Van de Ven, A., Poole, M.S.: Organization science. Special Issue: Longitudinal Field Research Methods for Studying Processes of Org. Change, vol. 1, no. 3, pp. 313–335 (1990)
11. Efindi, T.H.A.: The development of innovation capabilities. A review paper about the challenges and future research trend in the UAE high education. Test Eng. Manage. **83**, 7299–7307 (2020)
12. Andrusiv, U., Kinash, I., Cherchata, A., Polyanska, A., Dzoba, O., Tarasova, T., Lysak, H.: Experience and prospects of innovation development venture capital financing. Manage. Sci. Lett. **10**(4), 781–788 (2020)
13. Zmyzgova, T., Polyakova, E., Prokofyev, K., Chelovechkova, A., Dmitrieva, O.: University relations: university–industrial relations as the main factor in the development of polytechnic education. In: Solovev, D.B., Savaley, V.V., Bekker, A.T., Petukhov, V.I. (eds.) Proceeding of the International Science and Technology Conference "FarEastCon 2019". SIST, vol. 172, pp. 569–579. Springer, Singapore (2020)
14. Kirillova, E.: The role of scientific and industrial cooperation in assessing the innovative potential of an industrial enterprise and the approach to evaluation through joint patent and licensing activities. In: Solovev, D.B., Savaley, V.V., Bekker, A.T., Petukhov, V.I. (eds.) Proceeding of the International Science and Technology Conference "FarEastCon 2019". SIST, vol. 172, pp. 507–516. Springer, Singapore (2020)
15. Robertson, J., McCarthy, I.P., Pitt, L.: Leveraging social capital in university-industry knowledge transfer strategies: a comparative positioning framework. Knowl. Manage. Res. Pract. **17**(4), 461–472 (2019)
16. Thomas, L.D.W., Autio, E.: Innovation ecosystems. Oxford Research Encyclopaedia of Business and Management. Aldag, R. (ed). Oxford University Press, Oxford. (PDF) Innovation Ecosystems (forthcoming). https://www.researchgate.net/publication/337149363_Innovation_Ecosystems
17. Gunasekara, C.: Reframing the role of universities in the development of regional innovation systems. J. Technol. Trans. **31**, 101–113 (2006)

18. Suorsa, A.R., Svento, R., Lindfors, A.V., Huotari, M.-L.: Knowledge creation and interaction in an R&D project: the case of the energy weather forecast. J. Doc. **76**(1), 145–172 (2019)
19. Filippov, S., Mooi, H.: Innovation project management: a research agenda. J. Innov. Sustain. (2010). ISSN 2179-3565
20. Ghaben, R., Jaaron, A.: Assessing innovation practices in project management: the case of Palestinian construction projects. Int. J. Innov. Sci. Res. **17**(2), 451–465 (2015). ISSN 2351-8014
21. Shinoda, M., Nishioka, K., Mishima, A.: Systematization of the method of project management for education in university. Educational Alternatives, vol. 15 (2017). ISSN 1314-7277
22. May, C.C.M., Hwa, Y.E., Spowage, A.: Developing and evaluating a project management methodology (PMM) for university-industry collaborative projects. Product: Management and Development, vol. 9, pp. 121–135 (2011)
23. Fernandes, G., O'Sullivan, D., Pinto, E.B., Araújo, M., Machado, R.J.: Value of project management in university–industry R&D collaborations. Int. J. Manag. Projects Bus. **13**(4), 819–843 (2020)
24. Säisä, M.E.K., Tiura, K., Matikainen, R.: Agile project management in university-industry collaboration projects. Int. J. Inf. Tech. Project Manag. **10**(2), 8–15 (2019)
25. Cassanelli, A.N., Fernandez-Sanchez, G., Guiridlian, M.C.: Principal researcher and project manager: who should drive R&D projects? R&D Manage. **47**(2), 277–287 (2017)
26. Hansen, I.-E., Mork, O.J., Welo, T.: EXPLORING Framework for university-industry innovation projects: building collaborative knowledge platform. In: European Conference on KM (2019)
27. Kowang, T.O., Long, C.S., Rasli, A.: Innovation management and performance framework for research university in Malaysia. Int. Educ. Stud. **8**(6), 32–45 (2015)
28. Troshani, I., Rampersad, G., Plewa, C.: Organisational adoption of e-business: the case of an innovation management tool at a university and technology transfer office. Int. J. Network. Virt. Organ. **9**(3), 265–282 (2011)
29. McLaughlin, D., Kennedy, D.: Innovation Project Management Handbook. Productivity Press, New York (2016). https://doi.org/10.1201/b19726
30. Yordanova, Z., Stoimenov, N., Boyanova, O., Ivanchev, I.: The long way from science to innovation – a research approach for creating an innovation project methodology. In: Abramowicz, W., Corchuelo, R. (eds.) BIS 2019. LNBIP, vol. 353, pp. 371–380. Springer, Cham (2019). https://doi.org/10.1007/978-3-030-20485-3_29
31. Lill, W., Gleich, R.: Agility and the role of project — internal control systems for innovation project performance. Int. J. Innov. Manag. **24**(07), 2050064 (2020)

Knowledge-Based Practices in Disaster Management: Development Towards Digital Humanism

Pg Hj Asmali Pg Badarudin(✉) and Fahmi Ibrahim

Universiti Teknologi Brunei, Jalan Tungku Link, Gadong,
Bandar Seri Begawan, Brunei Darussalam
salamdamit@gmail.com, fahmi.ibrahim@utb.edu.bn

Abstract. While new innovative and disruptive technologies are obsoleting existing technologies and necessitating organizational reconfiguration and restructuring in many spheres of life and businesses, the field of disaster management particularly disaster mitigation, is also not isolated from the beneficial changes brought by the transformational capabilities of digitalization. Lack of inter-agency co-operation and co-ordination in disaster risk management, however, can be an obstacle towards this. The mitigation of disasters like flood requires high degree of collaboration among multi-agencies with specialized knowledge in different disciplines as they work towards the same purpose. This paper proposes a conceptual framework of digitized knowledge-based collaboration platform that emphasizes digital humanism which has the capabilities of transforming multi-agencies collaboration effectively, faster and at lower costs towards greater knowledge creation and better mutual understanding. The idea is to put people at the center of the risk reduction efforts. By doing so, the overall digitization picture would accommodate organizational culture and people perspective towards enabling KM mechanisms such as Communities of Practices (CoPs) and other knowledge-based practices that would provide effective interaction and collaboration among experts, practitioners and researchers. This paper contributes a guideline for practitioners not only to succeed in implementing knowledge-based disaster management but to secure the deliverables for public safety and greater community resilience in this digital age. Accordingly, the resultant framework is intended as a useful guideline for drawing attention to the theoretical underpinnings of the concept of Information Systems, Knowledge Management (KM), understanding the interplay between these and the aims of disaster management.

Keywords: Knowledge-based collaborative practices · Digital humanism · Disaster management

1 Introduction

Digital technologies are increasingly taking up new roles in human daily lives and at workplace, making tasks easier, faster and cheaper. The business and organizational

W. S. H. Suhaili et al. (Eds.): CIIS 2020, AISC 1321, pp. 13–25, 2021.
https://doi.org/10.1007/978-3-030-68133-3_2

world has experienced the emergence of new innovative experiences, services, organizations and even industries that have never been imagined before, disrupting incumbent industries and institutions, bankrupting irrelevant business organizations, obsoleting existing technologies and reconfiguring and restructuring orders of things and ways of doing things in almost every facet of the socio-economic sphere.

The field of disaster management, specifically in disaster mitigation is not isolated from these digital changes as it can also gain benefits from the transformational capabilities of digitalization. For instance, hydro-meteorological disasters in Brunei Darussalam have been increasing in numbers and intensity and becoming more unpredictable in the recent decades as a result of climate change, global warming or other phenomena including man-made human development. Despite advances in artificial intelligence (AI) particularly in predictive analytics, the development has made the tasks of forecasting and predicting the weather more challenging, adding new layer of complexity to an already complex field. It is known that disaster management professionals have been using sophisticated information and communications technologies ever since the emergence of early weather monitoring platforms either surface-based like radars and sensors or space based like remote-sensing satellites and airplanes and currently drones and Internet-of-Things (IoT) sensors. In the prediction of hydro-meteorological disaster like floods, flash floods or strong winds, multiple agencies collaboration are the norm. In the case of Brunei Darussalam where this study is being made, multiple agencies have to work together in order to gain or share knowledge, information and data for disaster mitigation purposes like prediction. Coordination with regional agencies in neighboring countries and further afield are also essential for the purpose of disaster recovery and relief, in tandem with regional and international agreements such as ASEAN Agreement on Disaster Management and Emergency Response (AADMER). With the convergence of digital devices and ICT platforms, particularly the cloud computing paradigms, coupled with the proliferation of mobile devices and embedded Internet of Things or IoT sensors, the increasing digitization of data has provided disaster management professionals and domain knowledge experts with new opportunities for increasing the collaborations on risk management through new digital channels.

Today, a tap on a screen can get a weather alert. A fisherman can simply cancel going to the sea due to a rough sea warning from an app. However, behind the screen, weather experts may be working continuously to provide greater accuracy of the weather warning algorithm as a result of changing climate and other phenomena that require greater and wider collaborations with other experts and professionals in multiple fields. A meteorological agency may today be providing historical rainfall data to a hydrology organization, who will then analyze the data for the purpose of designing a better drainage system. This collaboration can be improved by providing a digital platform that not only allows both agencies but may also allow agencies or subject matter experts with important specializations that can contribute in view of the evolving hydro-meteorological disaster risks and the impact of human development on exacerbating the disaster risks. In the eagerness to conform to the digitization agenda, agencies madly rush in the development of digital collaboration and platforms without considering the human aspects or behavior in respect to knowledge and expertise sharing. This impulsive response may ultimately lead to major digitization project failures costing thousands or even millions of dollars in collateral damage. Errors in certain sets of important data and their subsequent analysis may lead to ineffective investment in a country's drainage systems, for example. This paper attempts to address one aspect that may be overlooked that is the need for digital humanism in developing disaster mitigation collaboration platforms in flood risk management, using Brunei Darussalam as a case study.

2 Background

Hydro-meteorological disasters like floods, flash floods, strong winds, drought, wildfire and landslides are among the most common disasters worldwide. Floods have particularly posed the most hazardous risks that cause frequent heavy economic loss. In Brunei Darussalam, a country which is mostly spared from major disasters, floods had even claimed two lives in 2009. Every year, flooding in general is responsible for more deaths around the world than any other type of natural disaster [1]. Flash floods are particularly dangerous because of their signature trait: speed [2]. In Brunei Darussalam, flash floods are also more difficult to predict due to the combination of natural and anthropogenic factors related to human developmental actions [3] with rapid urban and rural development as well as the short overland time of movement of water especially in the district of Brunei-Muara compared to Tutong or other two districts (Kuala Belait and Temburong) where the rivers are longer and the flood water rising slowly. In view of these factors, flash floods require a lot of collaboration among subject matter experts and stakeholders in various fields and expertise. No single agency or institution has expert and specialized knowledge in the numerous issues related to hydro-meteorological disasters—from experts in many scientific fields such as meteorology, oceanography, geography to human development areas such as drainage engineers, town and country planners and many more—that can possibly work alone in the mitigation of flash floods. Therefore, an integrated and effective coordinated hydro-meteorological disaster management is strongly recommended, characterized by regular and frequent stakeholder consultations [3] with the addition of the digital humanism dimension.

3 Mechanisms Process of Collaboration in Hydro-meteorological Disaster Mitigation

Brunei Darussalam's hydro-meteorological disaster mitigation, is led by the Department of Drainage and Sewerage (DDS) as the lead agency while Brunei Darussalam Meteorological Department (BDMD) and National Disaster Management Centre (NDMC) which are important in providing data and coordination at national level respectively. Agencies and institutions such as Universiti Brunei Darussalam (UBD) and Universiti Teknologi Brunei (UTB) have researchers working on projects in disaster management. UBD has partnership with IBM which provides IBM Supercomputer in flood risk management research. Nevertheless, it was indicated that in Brunei there is a lack of inter-agency co-operation and co-ordination in disaster management (Flood Risk Management Conference, 2015)

With the declining costs of global communication and information processing, hierarchy is being replaced by radically different ways of organizing [4], the digital elements of which include cloud computing, big data analytics, cognitive computing, and collaboration platforms [5]. The current adoption of digital technologies for floods and flash floods for flood disaster mitigation agencies are shown in the Table 1 below. The table shows the consolidation description analysis of available Early Warning Systems (EWS) by the relevant agencies. The availability of database also is indicated in each of relevant agency. It is important to note that the challenge of collaborative arrangement of each agency is described in the Table 1 below. For example, there is no integration or inter-operable collaboration platform with BDMD or other agencies. This demonstrates the disaster management agencies' systems are functioning in silos without inter-operability and integration. As an alternative of EWS, other digital technologies for disaster mitigation are also investigated such Mobile App for public Weather called Brunei WX.

Table 1. Consolidation analysis of Early Warning Systems (EWS) Source: Authors.

Agencies	Early Warning Systems (EWS}	Database availability	Collaborative arrangement	Other digital technologies
DDS	• For DDS use only using IoT sensors at selected rivers basins. Since 90s DDS already implemented electronic EWS for Tutong river but has been recently out of order mostly due to lightning strikes • Digital and analogue or manual at certain areas have to request Fire and Rescue Department personnel to measure water level in various areas manually	None	• Request BDMD for rainfall data using normal communications • No integrated or inter-operable collaboration platform with BDMD or other agencies	• Website and Mobile Apps for monitoring of water levels and other relevant data from remote station • For DDS use only • No mobile app for public yet • No cloud usage

(*continued*)

Table 1. (*continued*)

Agencies	Early Warning Systems (EWS}	Database availability	Collaborative arrangement	Other digital technologies
BDMD	• For public using website and mobile app for weather information where forecasted heavy rainfalls or extended rainfall will come with a warning of floods or flashfloods in areas. The EWS are based on monitoring of Climate Database System • No integrated or inter-operable collaboration platform • Mobile App for public weather forecasts called Brunei WX 5 surrounding weather conditions current weather conditions involving scales of "mukim", district, nation and region surface and space-based weather observation systems [6]	Climate Database System	No integrated or inter-operable collaboration platform	Mobile App for public weather forecasts called Brunei WX

(*continued*)

Table 1. (*continued*)

Agencies	Early Warning Systems (EWS}	Database availability	Collaborative arrangement	Other digital technologies
NDMC	NDMC operates at the operational/national level and has coordinating functions among agencies in the tactical level	None	• Request DDS for flood hazard map • Collaboration with CAE MPTC on disaster simulation modeling and training [7] • No official receipt of researches done by research universities or researchers even though requests are made • No integrated or inter-operable collaboration platform with BMBD or other agencies	Use the regional ASEAN Disaster Monitoring and Response System (DMRS)
UBD	The UBD/IBM Centre increase warning times from the one hour to 18 hours. By feeding LIDAR (Light Detecting and Ranging) data into their state-of-the-art IBM Blue Gene supercomputer, the Centre can come up with a model that predicts which areas will experience flash flooding [8]		Available	• Undergraduate or graduate research projects require data from BDMD, DDS and other disasters agencies like Fire and Rescue Department • Ministry of Health • No integrated or inter-operable collaboration platform

(*continued*)

Table 1. (*continued*)

Agencies	Early Warning Systems (EWS}	Database availability	Collaborative arrangement	Other digital technologies
UTB	River water level using IoT project	None	• Undergraduate or graduate research projects require data from BDMD, DDS and other disasters agencies like Fire and Rescue Department and Ministry of Health • No integrated or inter-operable collaboration platform	

In sum, from the analysis Table 1 above, it shows there is no inter-agency collaboration using an integrated and inter-operable digital technology. Increasing discussions and debates within disaster mitigation group of agencies have raised questions regarding the practicality of adopting current ad-hoc collaborative approaches to disaster management, which ironically is akin to the setting up of ad-hoc disaster management committees prior to the enactment of the Disaster Management Order in 2006 [9]. This ad-hoc approaches to collaboration could easily lead to short-term and hasty decisions adversely affecting the agencies' ability to achieve sustainable long-term goal. To minimize the occurrence of these unwise decisions, it is important to plan proactively for disaster mitigation in order to provide general guidance for decision-makers and a framework for the professionals involved in disaster management processes. For proactive plans and decisions, a knowledge-based collaboration platform with digital humanism perspective is thus proposed given major issues relating to the risk management of this disaster require experts' knowledge solution.

It is suggested that a Knowledge Management (KM) approach may assist in improving collaboration processes and coordination because the most likely areas on which to focus to reduce unwise and ineffective decision can be identified during the early stage of the disaster scenario. Tapping on the past experiences of disaster scenarios, the KM approach provides a wealth of pertinent and useful information for decision makers and may eventually enhance collaborative ventures. By having the knowledge-base and a systematic way to make well-informed decisions, combined with the efficiency of agencies and strong coordination, disaster mitigation success should increase.

4 Organizational Analysis: Brunei Disaster Management

In developing a knowledge-based collaboration platform with digital humanism perspective, and organizational analysis is required to assess the feasibility of the implementation. From the organizational analysis, it illustrates that there are three (3) level of strategic planning process of Brunei Disaster Management (see Fig. 1). National Disaster Council (NDC) formulates strategic policies [10] which convenes regularly. Strategic level is an organization's process of defining its strategy, or direction, and making decisions on allocating its resources to pursue this strategy while NDMC is to co-ordinate assistance for disaster management and disaster operations, whether within or outside Brunei Darussalam [10]; This makes NDMC to be suitable for ownership of the proposed digitized collaboration platform for disaster management mitigation with KM approach, being the central coordinating agency at operational level. Operational is the process of linking strategic goals and objectives to tactical goals and objectives. Finally, the various tactical agencies under NDMC will then be able to work together and collaborate. Tactical level is short range planning emphasizing the current operations of various parts of the BDMD. In addition, national risk assessments are conducted by respective agencies at different level. Risk assessment for natural disasters is conducted regularly by the NDMC and all these risk assessments involve multi-sectoral approaches [11].

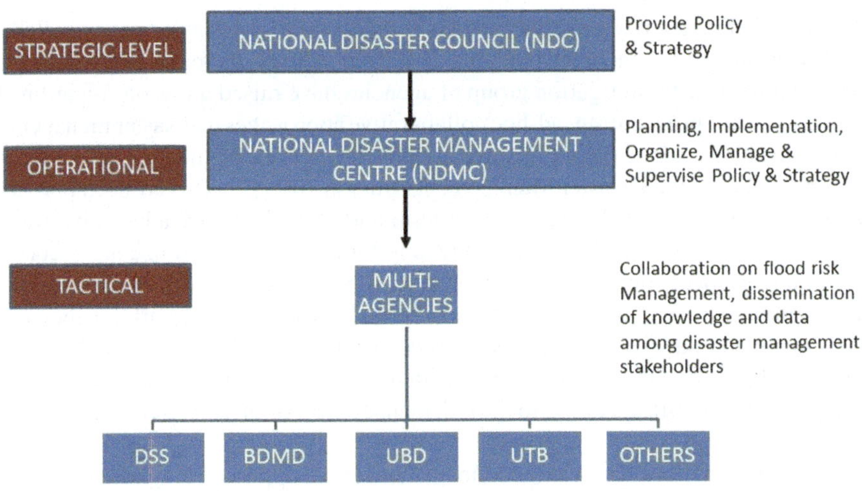

Fig. 1. Organizational analysis of Brunei Darussalam disaster management for hydro-meteorological disaster mitigation teams (Adapted from [12])

5 Digitized Knowledge-Based Collaboration with the Embedded Digital Humanism

In the context of the digital era today, the hybrid approach suggested by Ibrahim and Barr (2009) and Ibrahim and Reid (2010) to provide for a balanced strategy in terms of KM practices which is crucial for successful KM implementation [13, 14] has some similarities to the digital humanism, as defined by Gartner [15], where people are the central focus in the manifestation of digital businesses and digital workplaces. Businesses who embrace digital humanism use technology to redefine the way people achieve their goals and enable people to achieve things not previously possible [15]. Against the opposite of a digital humanist which is the digital machinist where the machinist will seek to automate everything and put the user, the people out of the equation as far as possible, and maybe even entirely [15]. Digital humanists' work lies at the cross-roads of computer science and the humanities [16]. Moving traditional humanistic material into digital forms brings new challenges for scholars of the humanities. The proposed Digitized Knowledge-based Collaboration is intended to help users make humanistic knowledge in digital media more manageable and usable.

This paper proposes components of a digitized knowledge-based collaboration with the embedded digital humanism as shown in Fig. 2 below. It is designed to be an integrated and interoperable collaboration platform that addresses the current lack of inter-agency cooperation and coordination in disaster mitigation while embedding digital humanism concept. The central idea is to ensure that the disaster mitigation team will always be involved actively in the adoption of digital technologies and development of digitalized collaboration platform from the beginning of the process with the technical development team – a collaboration called digital humanism development team This includes constant reference to the ultimate aims that the platform is being assigned to achieve and the lesser goals that help to achieve the ultimate aims of disaster mitigation such as reduced risks and vulnerabilities, improved disaster responses and improved disaster resilience. Figure 2 illustrates six main mechanisms described as follows as indicated by the numbered circles;

1. Community of Practices (CoP)

CoP is a group whose members regularly engage in sharing and learning, based on their common interests [17]. It is also a platform that can enable collective action and responsibility in the management and sharing of knowledge among experts across various agencies and disciplines. CoP leads to changes in behavioral which lead to greater sharing of knowledge which in turn will positively influence performance [17].

Badarudin, Wan and Phon-Amnuaisuk (2017) have proposed CoP in a hydro-meteorological disaster mitigation platform [18] and the Philippines disaster management which has setup CoP for disaster preparedness in 2018, where the CoP will facilitate continuous learning and capacity building on communications as aid, support the conduct of simulation exercise on coordinated communication strategy across all-inclusive sectors and provide the platform for documentation and sharing of good practices [19].

A digitized CoP will further add new value in terms of the benefits of being digital. The development of digitized CoP will bring along many experts from different disciplines and from different background and organizations. The knowledge may be digitized to form expert systems that can be stored in the databases and knowledgebase for later use by decision-makers. Specific skills and know-how can even be counted as intangible digital assets and can be obtained from the digitized CoP to be used during emergencies for further improvement of asset visibility and collaboration as suggested in [20].

It is evident that the organizations commonly had focused people involved in leveraging knowledge through the establishment of communities of practices (CoPs) and cross-functional groups. Although disaster management focuses on IT knowledge-based tools, human input is still required in producing solution of disaster mitigation. This seems to indicate that knowledge is non-codifiable. This view was identified in KM literature there being no such thing as completely explicit knowledge because all knowledge will remain tacit and therefore resistant to articulation and codification [21]. In other words, there will be an element of tacit knowledge which is still embodied in people. This means, from the perspective of humanism (people involvement), knowledge is viewed as dynamic and socially constructed. This is at odds with the conception of knowledge in the role of IT in KM practices. By incorporating digital humanism will provide greater understanding of natural phenomena or man-made factors that causes floods and flash flood and the creation of new knowledge that can mitigate the risks.

2. Group Decision Support Systems (GDSS)

A group decision-support system (GDSS) is a platform that facilitates collaborative decision-making. Decision makers may use knowledge obtained from KM systems such as CoP and expert or recommendation systems. The GDSS itself may be digitized and automated such as by using intelligent agents with real human decision-makers will always be consulted and involved in order to verify the accuracy and correctness of the decisions. Digital humanistic mechanism includes verification of people to provide a consensus for the digitized solutions.

3. Artificial Intelligence (AI) – Data Mining and other tools

Data mining and other AI tools, in turn, can enhance the usefulness of both CoPs and GDSS and significantly increase the degree of digitalization by providing an element of predictive analytic capabilities along with pattern and new knowledge discovery. The data mining methodology and automation of the process should also involve the concepts of digital humanism. The development of new models should be shared among experts in the relevant field and the existing models should also be verified by human experts. The algorithms themselves can always be checked for accuracy and optimization processes should always be embedded. The use of cognitive computing like natural language processing (NLP) can increase the usability of the platform such that the participants can talk naturally to the platform rather than having to type text. The incorporation of this AI tool to the knowledge-based platform should require digital humanistic input such that experts or specialist may speak about something which has different meaning or context in their respective fields in order to increase mutual understanding and remove biases and stereotyping.

4. Other Digital Tools i.e. Cloud and Mobile Computing and IoT

Besides lacking collaboration and interoperability, current framework also lacks inter-agency data collection of previous disasters, an issue often cited in local DM circles. Thus, input and output mechanisms that include real-time and online data capture, processing and visualization can be built-in into the CoP and GDSS and the collaboration platform. Cognitive computing mentioned earlier will improve the input process as well as it makes input faster better and more accurate.

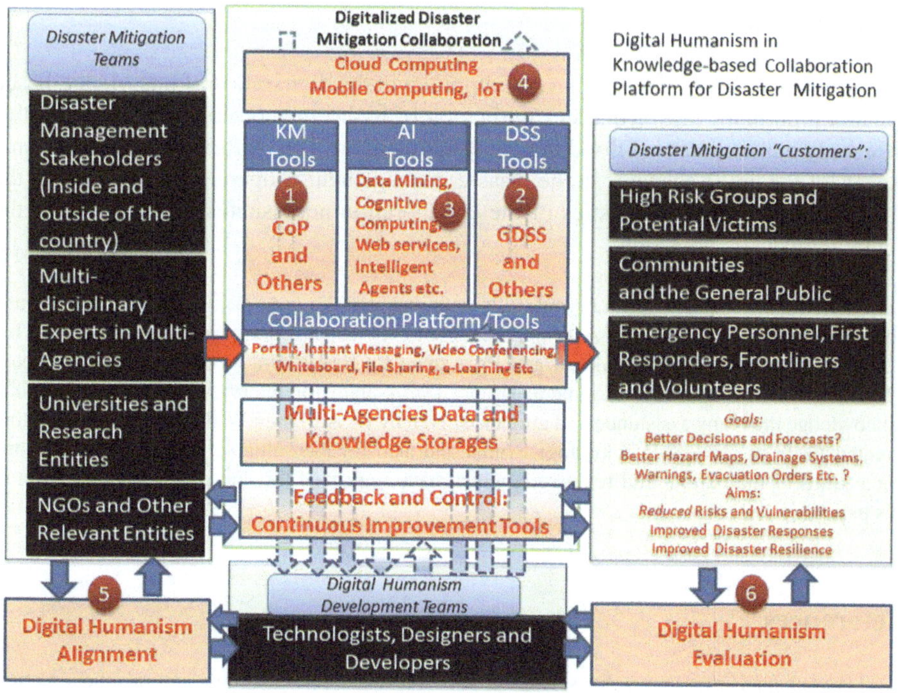

Fig. 2. Digitized knowledge-based collaboration with the embedded digital humanism. Source: Authors.

5. Digital Humanism Alignment

There should also be mechanisms for a so-called digital humanism alignment process among the disaster management stakeholders in general and the hydro-meteorological disaster mitigation teams in particular and the digital humanism development team to ensure that the goals of disaster management are always taken into account in any digitization project or program starting from the conceptualization and design stages until the implementation and maintenance stages (See Fig. 2).

6. Digital Humanism Evaluation

Lastly, the achievement of strategic aims of disaster management such as flood risk and vulnerabilities reduction and the improved disaster responses and resilience should continuously be measured and monitored through a systematic process of digital humanism evaluation. Questions like "Were better decisions and forecasts achieved?", "Is there a better hazard map utilization?", "Are there better drainage systems and early warning systems?" could trigger feedback for improving the performance of disaster mitigation and ultimately the whole disaster management (See Fig. 2).

6 Conclusion

The need for digital humanism is very important in the digitization of knowledge-based collaboration platform for disaster mitigation of hydro-meteorological disaster in this case study in the context of flood in Brunei Darussalam. The cultural and organizational aspects of the multi-agencies in the disaster management cooperation coupled with the multi-disciplinary characteristics of the collaboration necessitates the adoption of the digital humanistic approach.

Acknowledgements. The authors gratefully acknowledge the kind assistance given by Awang Shahrin bin Haji Mohd Jawie, Acting Director of NDMC and his officers Abdul Muein bin Adi Marhain and Mohd Adib Matali and Pg Azman Pg Badarudin, Acting Assistant Director, DDS, PWD, Ministry of Development, Brunei Darussalam for interviews. We also gratefully acknowledge the many assistance and guidance given by Dr Au Thien Wan and Dr Somnuk Phon-Amnuaisuk from Universiti Teknologi Brunei and also for the valuable helpful comments and suggestions of the editors and reviewers, which have improved the content and presentation of this paper.

References

1. Live Science. https://www.livescience.com/32410-whats-the-deadliest-natural-phenomenon.html. Accessed 01 Sept 2020
2. How Stuff Works. https://science.howstuffworks.com/nature/natural-disas-ters/why-flash-floods-are-so-dangerous.html. Accessed 01 Sept 2020
3. Ndah, A.B., Kumar, D.L., Becek, K.: Dynamics of hydro-meteorological disasters: revisiting the mechanisms and drivers of recurrent floods and landslides in Brunei Darussalam. Int. J. Earth Atmos. Sci. (2016)
4. Fjeldstad, Ø.D., Snow, C.C., Miles, R.E., Lettl, C.: The architecture of collaboration. Strat. Manage. J. **33**(6), 734–750 (2012)
5. Snow, C.C., Fjeldstad, Ø.D., Langer, A.M.: Designing the digital organization. J. Organ. Des. **6**(1), 1–13 (2017)

6. BDMD: Newly Empowered Brunei Darussalam Meteorological Department. http://www.mtic.gov.bn/Lists/News%20%20Press%20Release/NewItemDis-play.aspx?ID=141&ContentTypeId=0x0100D96C0410F8497A41ABB53106FFA4890D. Accessed 01 Sept 2020
7. Martin, J.D.: Brunei Disaster Management Reference Handbook. Center for Excellence in Disaster Management and Humanitarian Assistance, 456 Hornet Avenue, Joint Base Pearl Harbor, Hickam, Hawaii 96860-3503 (2018)
8. Flood Forecasting System – a high performance computing application. https://ubd.edu.bn/news/2014/03/18/flood-forecasting-system-%E2%80%93-a-high-per-formance-computing-application/. Accessed 01 Sept 2020
9. NIDM: National Institute of Disaster Management India. http://nidm.gov.in/easindia2014/err/pdf/country_profile/brunei_darussalam.pdf. Accessed 01 Sept 2020
10. Brunei Darussalam Government. Brunei Disaster Management Order. 2 March 2006. http://extwprlegs1.fao.org/docs/pdf/bru87653.pdf. Accessed 01 Sept 2020
11. Joint external evaluation of IHR core capacities of Brunei Darussalam. https://www.who.int/publications/i/item/9789240006973. Accessed 01 Sept 2020
12. National Institute of Disaster Management (NIDM): Presentation of Director, NDMC, Brunei Darussalam, June 2010. http://nidm.gov.in/easindia2014/err/pdf/country_profile/brunei_darussalam.pdf. Accessed 01 Sept 2020
13. Ibrahim, F., Barr, S.: Reconfiguring IT and the cultural dimension in knowledge management practices. Int. J. Knowl. Culture Change Manag. (IJKCCM) **9**(11), 27–43 (2009)
14. Ibrahim, F., Reid, V.: Integrated use of information technology and people involvement for knowledge management. Int. J. Technol. Knowl. Soc. (IJTKS) **6**(2), 163–180 (2010)
15. Embracing Digital Humanism. https://www.gartner.com/smarterwithgartner/embracing-digital-humanism/. Accessed 01 Sept 2020
16. Stanford Humanities Center Digital Humanities. http://shc.stanford.edu/digital-humanitis. Accessed 01 Sept 2020
17. Lesser, L.E., Storck, J.: Communities of practice and organizational performance. IBM ~ Syst. J. **40**, 831–841 (2011)
18. Badarudin, P.H.A., Wan, A.T., Phon-Amnuaisuk, S.: A knowledge-based framework for mitigating hydro-meteorological disasters. In: DMBD 2017, Fukuoka, Japan, pp. 505–513. Springer, Cham (2017)
19. ReliefWeb Philippines: Community of Practice (CoP) on Community Engagement Preparedness and Response Terms of Reference (ToR) (2018). https://reliefweb.int/report/philippines/philippines-community-practice-cop-community-engagement-preparedness-and-0. Accessed 01 Sept 2020
20. Badarudin, P.H.A., Wan, A.T., Phon-Amnuaisuk, S.: A blockchain-based assistance digital model for first responders and emergency volunteers in disaster response and recovery. In: 8th International Conference on Information and Communication Technology (ICoICT), Yogyakarta, Indonesia, pp. 1–5. IEEE (2020)
21. Newell, S., Robertson, M., Scarbrough, H., Swan, J.: Managing Knowledge Work. Palgrave Macmillan, New York (2002)

Destructive Digital Ecosystem of Cyber Bullying Perfective Within the Information Technology Age

Desi Setiana[1,2](✉) and Norainna Besar[2]

[1] Ministry of Law and Human Rights Republic of Indonesia, Jakarta, Indonesia
desi.setiana@gmail.com
[2] University of Brunei Darussalam, Bandar Seri Begawan, Brunei
norainna.besar@ubd.edu.bn

Abstract. With the rapid growth of modern technologies, it is significant to address and understand the issue of cyberbullying and its consequences. Therefore the purpose of this paper is to synthesise the literature on cyberbullying among students, but with particular reference to cyberbullying as an aggressive act which causes criminal behaviour and affects students' life. The study is based on the analysis of a large number of publications on cyberbullying (books, journal articles, and conference papers). Analysis of the literature suggests that cyberbullying is a perilous act which could give devastating consequences to the victims specifically to students who are affected as they are still enrolling in their schools. Moreover, cyberbullying can affect a victim's life day and night without stopping, unlike traditional bullying, as soon as the student reaches home, the activities stop. However, whether traditional bullying or cyberbullying, suffer emotionally and psychologically. Support from family, friends, schools, health care providers play an essential role in helping students overcome cyberbullying and negative after-effects of cyberbullying.

Keywords: Information Technology Age · Cyber Bullying · Digital Ecosystem · Cyber Victims · Criminal Behavior

1 Introduction

The emergence of portable electronic devices with internet access has become essential to our lives as we heavily depend on it. In this era, the internet can be accessed quickly. The Internet has become essential to our lives as it provides us with an abundance of information which is needed to enhance our knowledge, educate each other and stay up to date with news. The Internet is also being used to socialise and as a medium to connect with our family and friends, particularly in social media platforms anytime, anywhere. Although modern technology gives us plenty of advantages, it also comes with negative consequences, as an example-cyberbullying.

Cyberbullies may lie on the basis that the victim may reveal his/her personal details or information that can be used against him/her. Hacking someone's email or online

W. S. H. Suhaili et al. (Eds.): CIIS 2020, AISC 1321, pp. 26–32, 2021.
https://doi.org/10.1007/978-3-030-68133-3_3

account, such as an online bank account can also be one of the activities that bullies can do for the person to be embarrassed or even spreading false information. Cyberbullying represents an aggressive form of communication that utilises electronic medium which will have an impact on the victim's mental health conditions. Moreover, this study explores concerning three main questions as follows; "What is the cause of cyberbullying among students?" "How do cyberbullying affects students?" "What are students perceptions of cyberbullying?".

2 Research Methodology

In this paper, the literature review as a research method is conducted. A literature review can broadly be described as a less systematic way of collecting and synthesising previous research by integrating findings and perspectives from existing published studies and records [14]. Therefore, the purpose of this paper is to synthesise the literature on cyberbullying among students, but with particular reference to cyberbullying as an aggressive act which causes criminal behaviour and affects students' life.

To establish the list of publications to be reviewed, The following keywords to search the literature: Information Technology Age, Cyber Bullying, Digital Ecosystem, Cyber Victims, Criminal Behavior were used. To include the disciplines of social studies, education and anthropology, we used the following databases: ELSEVIER, SAGE, Google Scholar and JSTOR within 9 years back. We limited our findings to articles, books, book chapters, websites that provided information about one or more of the following: (1) The Forms of Cyberbullying; (2) The Causes of Cyberbullying; (3) The Effects of Cyberbullying; (4) The Perceptions of Cyberbullying.

3 Discussion

3.1 The Forms of Cyberbullying

"Cyberbullying" occurs when a person or a group of people use electronic devices such as mobile phones, emails, social networks, blogs and web pages to bully other people intentionally and repeatedly such as by giving insults, attacking them and other means to hurt them [2].

There are various forms of cyberbullying [2]: *Harassment*: the bullies keep insults and offensive messages repeatedly; *Cyberstalking*: victims keep receiving threats and intimidating messages; *Slandering*: Bullies spread rumours and gossip intending to ruin other people's reputations; *Breach of privacy*: spread out someone's personal information without their consent to the public; *Password theft*: hack into someone's account to cause them harm; *Exclusion*: intentionally leave out someone in an online group.

There is another issue that could lead to an increase of cyberbullying, which is the lack of involvement of the school regarding the issue. Based on the study [3], they mentioned that though they are aware of the issue, they do not see it as a severe problem in their schools.

Due to the increasing number of bullying, the involvement of parents, school staff, children and other professionals are very encouraged to reduce it. Therefore parents and adults should increase their knowledge regarding new technologies to supervise the youths and their activities in the cyberspace [2]. The parents are also the one who is responsible for monitoring their children's behaviours outside the school [1]. Based on another study, many youths stated the fear of their parents knowing about them being involved in cyberbullying or their cyber victimisation as it could limit their access to technologies which refrains them from interacting in the cyberspace [7].

It was seen that most youths have lack of knowledge regarding cyberbullying and the importance of basic e-security applications in the cyberspace [15–19]. Cyberbullying affect both the cyberbullies and cyber victims, but it also affects the bystanders [21]. Bystanders are those who witness the cyberbullying in which they can either give support to the victims anonymously or asynchronously. They also stated that when the bystanders decided to intervene, it is to help the victims after the cyberbullying.

Support from family, friends and teachers play a significant role in reducing the adverse effects caused by bullying [1]. Those who have a high level of support from their family and friends will decrease the possibility of them being cyberbullied. The parents can provide social support to their children by listening to their problems and concerns, help them with their struggles and just show them that they genuinely care. These could help develop their sensitivity towards cyberbullying. Health care providers (HCPs), who consist of physicians, social workers, nurses and also psychologists could also help in reducing cyberbullying by educating the youths [9]. It was also proven that cyber victims do want their support. Health care providers also have the opportunities for collaborations with the school to make better social relationships as well as provide support and educate the teachers on how to handle the cyberbullying issues.

3.2 The Causes of Cyberbullying

After analysing the data regarding each research question, some relations were identified and indicating that a majority of respondents were a victim of cyberbullying especially during their school years. One out of every five (20.2%) students are reported to be bullied. The research indicates that a higher percentage of male than of female students report being physically bullied (6% vs 4%). In contrast, a higher percentage of female than of male students reported being the subjects of rumours (18% vs 9%) and being excluded from activities on purpose (7% vs 4%). Of those students who reported being bullied, 13% were made fun of, called names, or insulted; 13% were the subject of rumours; 5% were pushed, shoved, tripped, or spit on; and 5% were excluded from activities on purpose [10]. Bullied students reported that bullying occurred in the following places: the hallway or stairwell at school (43%), inside the classroom (42%), in the cafeteria (27%), outside on school grounds (22%), online or by text (15%), in the bathroom or locker room (12%), and on the school bus (8%) [10]. Reported 23% of African-American students, 23% of Caucasian students, 16% of Hispanic students, and 7% of Asian students report being bullied at school [10].

There are numerous reasons why cyberbullying occurs or what motivates an individual to resort to cyberbullying. In contrast to traditional bullying, cyberbullying eliminates the physical requirement to act. This means the absence of face-to-face exchange of hateful speech towards one individual motivates people to cyberbully. This is mainly due to most cyberbullies considered themselves as "invisible" and cannot be detected to face their consequences [5].

There is a direct correlation between parental autonomy support and adolescent's overall behaviour. The term 'parental autonomy support' can be defined as supportive actions provided by parents to their children to help in nurturing and developing their inner motivational resources by acknowledging the adolescent perspectives and offering supportive ideas. Here is an indirect connection between parental autonomy support and cyberbullying that further proposes how adolescents behave when the need for autonomy is provided. A satisfied adolescent, in this case, student, will be able to determine the differences between good and destructive behaviours, acknowledge various emotions and develop empathy towards other people [4].

3.3 The Effects of Cyberbullying

There are many effects or consequences of cyberbullying. The effects of cyberbullying led to depression, suicidal ideation, drug abuse and alcoholism. The previous study has shown a higher level of depression and suicidal ideation and an increase in emotional distress, externalised hostility, and delinquency revealed by victims of cyberbullying as compared to non-victim. It is shown that the level of depression is associated with the level and degree of cyberbullying faced by the victim [13]. There is a positive relationship between depression and alcohol use and thus, heavy drinker led to high-risk factor for suicide [13]. Previous studies proved that there is a significant relationship between victim in cyberbullying and emotional disorder, the results point out that there is a significant relationship between cyber victimisation and depression among young adults [11].

The effects of cyberbullying led to a health problem on cyber victims. The cyber victims were distraught due to online harassment [8]. In comparison to traditional bullying, cyberbullying victims are little to less likely to know their bully; thus, the increase in their fear [11]. The bully may be among their circle, and there is an increased fear of victimisation related to cyberbullying victimization [12].

The effects of cyberbullying on cyberbullies/perpetrators. A study conducted by Selkie [13] had found that the participants had drinking problems when they had experienced cyberbullying as perpetrators. Another survey [6] has assessed that both cyber victim and cyberbully experienced psychological disorder such as anxiety, depression and suicidal ideation and physical health problems such as headache, insomnia, stomach ache and skin problem.

The effects of cyberbullying in academic performance. Cyberbullying and cyber harassment have a negative influence on academic achievement. There is a relation between physical, psychological, and academic to both traditional bullying and cyberbullying [6]. Similarly, there were possible adverse effects between traditional bullying and cyberbullying on which if the cyberbully and victims were male participants. The male participants reported they experienced malicious behaviour physically and psychologically as well as academically due to the effect of cyberbullying [6].

3.4 Students' Opinions Regarding Cyberbullying

Most students are unaware of the significant impact cyberbullying can cause towards the victims, but they are aware of the presence of cyberbullying on the internet, the teasing and mocking to be just playful gestures that are to be expected across all social media sites. Students participating in the online community, specifically the online gaming community, are more than ever exposed to a higher chance of cyberbullying in the shape of verbal harassment. Online gamers are more than likely to cast their frustrations by throwing tantrums and crust languages to mock their teammates at any given circumstances. According to a research done by Usta in 2017, the most typical variation of cyberbullying behaviours is usually through the conduct of "special tricks" as well as "insulting/sexual/threatening social interactions in online games". Besides, a considerable number of famous online personnel are indirectly promoting the act of cyberbullying merely as a typical social interaction that would have little to no impact towards others. Mediums such as Youtube and Twitter are filled with creators/contributors that are capitalising on their audience's self-satisfaction to behave negatively towards an individual in which these behaviours are considered acts of cyberbullying. This, in turn, creates an understanding to adolescents and other related parties to treat cyberbullying with less concern, thus practising it with no hesitation.

Extra efforts from other parties surrounding the student. Despite the perception of not being able to do anything, some students also suggested that collective efforts need to be executed to minimise the act of cyberbullying especially at schools effectively. It is essential for the school, the family and the students themselves to admit responsibilities. Adequate information and attention from teachers and family respectively will also result in better online behaviours for the students [20]. By providing a supportive environment for the student to develop and evolve mentally, numerous anti-social behaviours can be prevented and this will help eradicate the cyberbully and aid cyber victims in recovering from their experiences. Here, schools need to have a set of policy or standardised regulations concerning online conduct between students. This will minimise the need for the school to respond in the event of cyberbullying occurring outside of the school.

Identifying the culprits will be less complicated when students within the school are encouraged to report incidents relating to cyberbullying. Communication between students and teachers/lecturers is expected to be convenient and straightforward. By giving the student a simple method to report, cyberbullying victims and culprits can be identified and be taken into action.

4 Conclusion

With the rapid growth of modern technologies in this generation, it is crucial, especially for the youths to understand the issue regarding cyberbullying and its consequences. Parents and the schools play an essential role in helping the youths overcome cyberbullying and the negative after-effects in terms of academics performances and psychosocial problems. Both the parents and schools should also take serious measures to monitor the youths and their activities in the cyberspace. The roles of bystanders and health care providers are also important.

In contrast, health care providers can educate and spread awareness regarding cyberbullying which could help to prevent or stop it from happening. Cyberbullying could affect students' embarrassment to self-harm, depression and anxiety. Cyberbullying regarded as a mere playful gesture, and this had resulted in conflict towards gamers, vloggers or YouTubers because they thought it was only a fun thing. Still, different people have different views of it. To maximise one self-satisfaction could as well giving others negative impacts and this is also called cyberbullying by using social media and video games.

References

1. Akturk, A.O.: Analysis of cyberbullying sensitivity levels of high school students and their perceived social support levels. Interact. Technol. Smart Educ. **12**(1), 44–61 (2015)
2. Buelga, S., Cava, M.J., Musitu, G., Torralbu, E.: Cyberbullying aggressors among Spanish secondary education student: an exploratory study. Interact. Technol. Smart Educ. **12**(2), 100–115 (2015). https://doi.org/10.1108/ITSE-08-2014-0025
3. Espelage, D.L., Jun, S.H.: Cyberbullying prevention and intervention efforts: current knowledge and future directions. Can. J. Psychiatry **62**(6), 374–380 (2013)
4. Fousiani, K., Dimitropoulos, P., Michaelides, M.P., Petergem, S.V.: Perceived parenting and adolescent cyber-bullying: examining the intervening role of autonomy and relatedness need satisfaction, empathic concern and recognition of humanness. J. Child Family Stud., 2120–2129. https://doi.org/10.1007/s10826-016-0401-1(2016)
5. Hinduja, S., Patchin, J.W.: Cyberbullying Identification, Prevention, and Response. Cyberbullying Research Center (2014). (www.cyberbullying.us)
6. Kowalski, R.M., Limber, S.P.: Psychological, physical, and academic correlates of cyberbullying and traditional bullying. J. Adolesc. Health Official Publ. Soc. Adolesc. Med. **53**(1 Suppl), S13–S20 (2013)
7. Meter, D.J., Bauman, S.: Moral disengagement about cyberbullying and parental monitoring: effects on traditional bullying and victimization via cyberbullying involvement. J. Early Adolesc. **38**(3), 303–326 (2018). https://doi.org/10.1177/0272431616670752
8. Mitchell, K.J., Finkelhor, D., Jones, L.M., Wolak, J.: Prevalence and characteristics of youth sexting: a national study. Paediatrics **129**(1), 13–20 (2012). https://doi.org/10.1542/peds.2011-1730
9. Moreno, M.A., Vaillancourt, T.: The role of health care providers in cyberbullying. Can. J. Psychiatry **62**(6), 364–367 (2017). https://doi.org/10.1177/0706743716684792
10. National Center for Educational Statistics. Student reports of bullying: Results from the 2017 School Crime Supplement to the National Victimization Survey. US Department of Education (2019). https://nces.ed.gov/pubsearch/pubsinfo.asp?pubid=2015056
11. Nixon, C.L.: Current perspectives: the impact of cyberbullying on adolescent health. Adolesc. Health Med. Ther. **5**, 143–158 (2014). https://doi.org/10.2147/AHMT.S36456
12. Randa, R.: The influence of the cyber-social environment on fear of victimisation: cyberbullying and school. Secur. J. **26**(4) (2013). https://doi.org/10.1057/sj.2013.22
13. Selkie, E.M., Kota, R., Chan, Y.-F., Moreno, M.: Cyberbullying, depression, and problem alcohol use in female college students: a multisite study. Cyberpsychol. Behav. Soc. Netw. **18**(2), 79–86 (2015). https://doi.org/10.1089/cyber.2014.0371
14. Snyder, H.: Literature review as a research methodology: an overview and guidelines. J. Bus. Res. **104**, 333–339 (2019)

15. Susanto, H., Caesarendra, W., Glowacz, A., Leu, F., Haghi, P.K., Khusni, U., Glowacz, A.: Managing cloud intelligent systems over digital ecosystems: revealing emerging app technology in the time of the COVID19 pandemic. Appl. Syst. Innov **3**, 37 (2020). https://doi.org/10.3390/asi3030037
16. Susanto, H., Kang, C.C., Leu, F.Y.: Revealing the role of ICT for business core redesign. SSRN Electron. J. https://doi.org/10.2139/ssrn.2734917
17. Susanto, H., Almunawar, M.N.: Information Security Management Systems: A Novel Framework and Software as a Tool for Compliance with Information Security Standard. CRC Press, Boca Raton (2018)
18. Susanto, H., Almunawar, M.N.: Security and privacy issues in cloud-based e-government. In: Advances in Electronic Government, Digital Divide, and Regional Development, pp. 292–321. IGI Global, Hershey (2016)
19. Susanto, H., Almunawar, M.N.: Managing compliance with an information security management standard. In: Encyclopedia of Information Science and Technology, 3rd edn., pp. 1452–1463. IGI Global, Hershey (2015)
20. Usta, E.: University students' views about their cyberbullying behaviors and self-exposition. J. Educ. Sci. **8**(22), 70–71 (2017). https://files.eric.ed.gov/fulltext/ED577247.pdf
21. Young, R., Miles, S., Alhabash, S.: Attacks by anos: a content analysis of aggressive posts, victim responses, and bystander interventions on a social media site. Soc. Media+ Soc., 1–14 (2018)

A Mobile Application for Mental Health Care During COVID-19 Pandemic: Development and Usability Evaluation with System Usability Scale

Muhammad Nazrul Islam(✉), Shahriar Rahman Khan, Noor Nafiz Islam, Md. Rezwan-A-Rownok, Syed Rohit Zaman, and Samiha Raisa Zaman

Department of Computer Science and Engineering, Military Institute of Science and Technology, Mirpur Cantonment, Dhaka 1216, Bangladesh
nazrul@cse.mist.ac.bd, shahriarkhan.ndc@gmail.com

Abstract. The COVID-19 pandemic has an alarming consequence on the mental health of individuals of the affected countries. Though, several digital solutions are developed to fight the COVID-19 pandemic, little attention has been paid to develop any useful digital solution for providing mental support, especially in the context of Bangladesh. Therefore, the objectives of this research are to elicit the primary user requirements for developing a digital solution; and to design and develop a mobile-based solution for providing mental health support to the people of Bangladesh. To achieve these objectives, firstly, a requirement elicitation study was conducted with 37 participants through semi-structured interviews to reveal the requirements for developing a mobile application. Secondly, a mobile application named 'Muktomon' [open one's mind] was designed and developed that includes features like provide virtual mental health assistance using artificial intelligence (AI) based chat bot, virtual therapy through videos and audios, support to establish communication with doctors for getting the mental health services, and a list of authentic news sources. Finally, the usability and usefulness of this application were evaluated through the System Usability Scale (SUS). The findings of the evaluation study showed that the application is usable and useful to provide a positive impact towards improving the mental health state during the COVID-19 pandemic.

Keywords: Mobile application · Mental health · COVID-19 · Chat bot · Usability · Artificial Intelligence (AI) · System Usability Scale (SUS)

1 Introduction

The COVID-19 was first identified in Wuhan, China in December 2019, while the first three confirmed cases of COVID-19 patients were detected in Bangladesh on 8 March 2020 [1]. This pandemic has caused social, economical, political, and

W. S. H. Suhaili et al. (Eds.): CIIS 2020, AISC 1321, pp. 33–42, 2021.
https://doi.org/10.1007/978-3-030-68133-3_4

public health crisis worldwide including in Bangladesh [2,3]. The government of Bangladesh has taken a number of initiatives to fight with COVID-19 including awareness initiatives through mass media, printed media, and online platforms; provide education through mass media and the internet; launch AI and big data-based prediction system; giving financial support to small and medium level of industries and the likes [4,5,22]. The COVID-19 pandemic noticeably impacts on the productivity of the industries, agricultural sectors, education, earning foreign exchange, and daily wage earners [6,7]. Extreme social isolation and losing jobs (or unemployment) are the major reasons for developing psychological problems irrespective of people's age group [8]. A few recent studies highlighted the mental health issues due to the COVID-19 pandemic in Bangladesh. For example, Bodrud-Doza [9] highlighted psychosocial and socio-economic insecurity among people due to the loss of lives and livelihoods during the pandemic, while Yeasmin [10] explored the impact of the COVID-19 pandemic on the mental health of children in Bangladesh. However, these studies suggest little to no solution to overcome mental health problems. As mental health issues are on spike [11,17], further research and development are required to understand the supporting measures for providing an effective digital solution to support the people not just physically but also mentally during the COVID-19 pandemic. Again, mobile technologies are expanding increasingly in Bangladesh [12,13], the use of mobile and smartphone technologies for providing health services are believed to provide quality health services.

Therefore, the objective of this research are to understand the fundamental requirements of developing a digital solution for providing mental health support and to develop a mobile-based solution for enhancing self-confidence and mental health to fight with COVID-19 pandemic in context of Bangladesh. The contribution of this research is two folded. At first, the functionality, design, and psychological components necessary to provide the most benefit to mental health support seekers are discovered; and then basing on the requirements, a usable application is developed in context of Bangladesh.

The rest of the article is organized as follows. Section 2 presents related works. The participant's profile, study procedure, and findings of requirement elicitation study are discussed in Sect. 3. Next, in Sect. 4, design and development of this system are illustrated briefly. The evaluation study with results are presented in Sect. 5. Finally, Sect. 6 contains conclusion and future expansion.

2 Related Work

A limited number of researches have been conducted focusing on the development of the assistive technologies for supporting the mental health patients, as well as the mass population to ease their daily lives and activities during COVID-19 pandemic. This section briefly introduces these work.

Rajkumar [14] reviewed 28 articles focusing to COVID-19 and mental health; and found that anxiety and depression are very common psychological reactions, which are associated with disturbed sleep during the COVID-19 pandemic.

Again, Khattar [15], explored the day to day living, activities, learning styles, and mental health of young students of India, during the unprecedented time through survey; and investigated how they were adapting to the new e-learning system while managing their social lives. In [16], a chatbot is proposed for mental healthcare which assists psychiatric for counseling by communicating with a user through dialogues.

A few studies have been conducted in the context of Bangladesh focusing on mental health during the COVID-19 pandemic. Banna et al. [11] highlighted the impact of the COVID-19 pandemic on mental health in Bangladesh and also found that sociodemographic factors are associated with mental health. Hossain et al. [17] focused on the mental health challenges suffered by the nation, and proposed strategies to overcome such challenges. Yeasmin et al. [10] found that a noticeable number of children are suffering from mental health issues, during the period of lockdown.

Apart from these, Teles et al. [18] reviewed the existing mobile applications (n = 216) developed for depression assistance and found that most of the apps are limited to English, Spanish, or Portuguese; while Islam et al. [19] reviewed the applications developed for fighting with COVID-19 pandemic and found that a very limited applications (n = 7) focused to the mental health support.

To sum up, a very limited number of studies were carried out on the digital solution for providing mental health assistance, especially for the people in Bangladesh. Moreover, only a few research discussed the design, development, and evaluation processes of any mobile application that provides mental health support. Thus, this research focused on this issue in context of Bangladesh.

3 Requirements Elicitation Study

The purpose of requirements elicitation study was to reveal the primary requirements to develop any mobile application or digital solution for providing mental health assistance during COVID-19 pandemic.

3.1 Participants Profiles

The interviews were conducted with a total of 37 participants (21 male and 16 female) with an average age of 30 ± 4.5 years. Among them, 14 were undergraduate students, 17 serving at different government and non-government organizations; and the remaining six were unemployed since starting the pandemic. During the pandemic period, 25 were staying at home where 12 were working outside the home.

3.2 Study Procedure

The study was conducted following the semi-structured interviews. A written consent was taken from each participant to ensure anonymity and confidentiality, while the Research and Development wing of the authors' institute provided the ethical approval for this research. During interviews, the interviewees were asked about their mental or emotional experience that they felt during the COVID-19 pandemic. The example set of questionnaires include *"Do you feel the need to talk to someone while being home quarantined?"* , *"Do you keep yourself updated with the COVID-19 situation of Bangladesh?", "How do you feel about not being able to meet with your friends and relatives during this pandemic?", "How did you feel when you heard that someone from your family got sick during this pandemic?", "How difficult it was to go to a hospital or doctor?"* Each interview was audio recorded and lasted around 12–16 min. The audio-recordings were later transcribed for analysis.

3.3 Revealed Requirements

The transcribed data were analyzed through a qualitative data analysis approach. As outcomes, the following requirements were revealed:

(a) *Provide mental health therapy:* One of the most interesting and important theme, that emerged from the analysis was the need for providing virtual therapy. 13 out of 37 (35%) participants stated about this issue. As people are confined to their residences for a long period of time due to lockdown, it is important to keep their mental health fit for a healthy life. The participants opined that it would be better to have a system where a proper lifestyle and mental health exercises would be suggested through video and audio calls by health professionals, as a form of virtual therapy.

(b) *Provide mental health assistance:* More than two-third (70%) participants stated about this concern. People, who are maintaining quarantine during this COVID-19 pandemic, are becoming self-centered and suffering loneliness. One participant stated that, *"I would like to talk or chat with someone who would listen and respond to me"*.

(c) *Establish communication between doctors and patients:* Since, most of the people cannot visit doctors or hospitals during the lockdown period, 28 (75%) participants agreed that a communication bridge is required between doctors and patients. Thus they felt the necessity of a system to communicate with doctors for reducing the risk of going out and being affected by COVID-19.

(d) *Provide authenticate information:* As people are locked down, so the only way to know what is going on around the world is through news sources and social media. Participants stated that they become confused and mentally weak as a lot of fake news are there on social media. More than 50% (19 out of 37) participants pointed out the necessity of authentic information sources on COVID-19.

(e) *Develop usable and secure digital solution:* Participants opined that any digital solution developed for supporting mental health should be user friendly as this application will be used by people having different levels of familiarity with smartphone. A total of 30 (81%) participants were concerned about the security, usability, and privacy issues of the digital solutions or applications that could be developed for providing mental health assistance.

4 Design and Development

Based on the findings of the requirement elicitation study, a conceptual framework for developing a mobile application is proposed as shown in Fig. 1. The proposed framework consisting of four modules including *self-care, chatbot, call for doctor* and *authentic source of information* to address the stated requirements for providing mental health support.

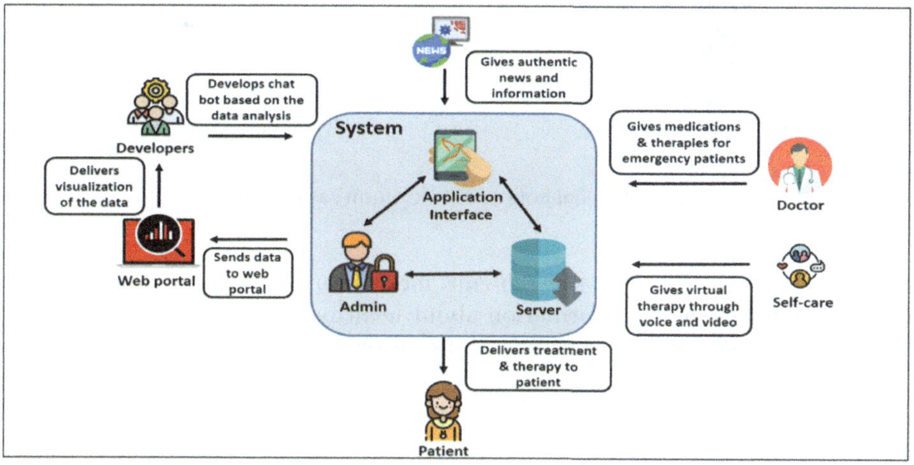

Fig. 1. Conceptual framework of the proposed application

The *self-care* module will facilitate user or mental health patients to take virtual therapy through voice and video services to mitigate mental health issues. The *chatbot* module will help the user to find out if any users have any mental health issues like stress, depression, anxiety, or loneliness through textual conversations. The textual conversation may include how the user is feeling and the chatbot will respond based on the user's queries or responses. Depending on the severity of the mental health condition as measured by the chatbot, the application may suggest seeking help from a therapist or to take self-care services through *self-care* module. The *call for doctor* module acts as a communication bridge between the doctors and the users. A user will find doctors in this module, who are ready to give services. Finally, in the *source of information* module, anyone can find the sources of real and authentic information related to COVID-19 to reduce the possibilities of being confused due to the fake news.

A mobile application named 'Muktomon' [in English, open one's mind] was developed based on the proposed conceptual framework. Android studio was used to design the user interface (UI). The application has four main sections: self-care, chatbot, call for doctor, and authentic source of information. A few screenshots of the developed application are shown in Fig. 2.

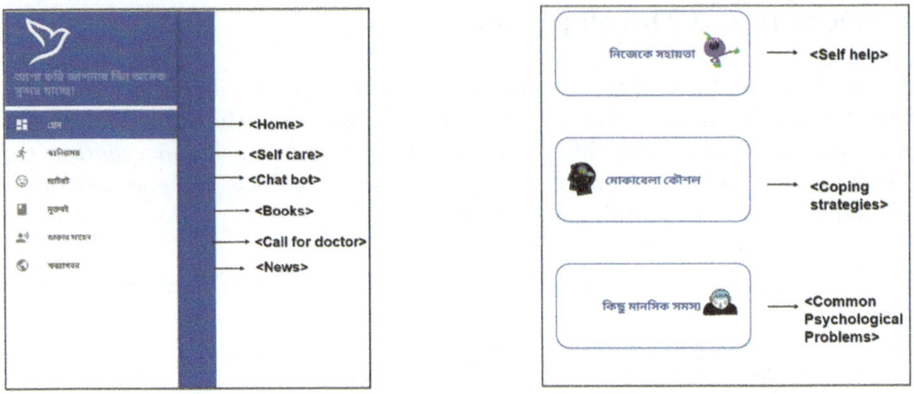

home page self-care interface

Fig. 2. Screenshots of ‘Muktomon’ application

The services included in the *self-care* module are virtual therapies through audios and videos; useful information about healthy routines and exercises; and information about meditations and mental disorders.

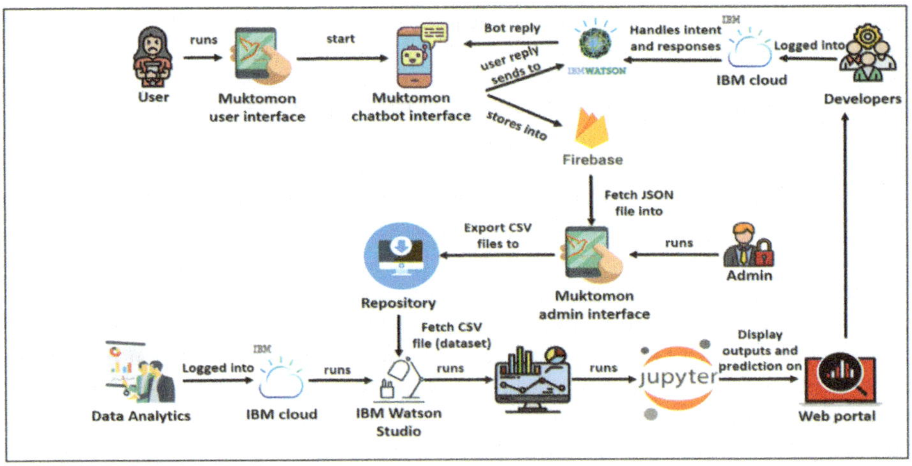

Fig. 3. Data flow of chat bot

The implementation process of the chatbot is shown in Fig. 3. The IBM Watson assistant was used to develop the chat bot [20]. In the IBM cloud, using the IBM Watson assistant, user’s and chat bot’s responses are handled by adding proper intents and dialogs respectively. As shown in Fig. 3, when a user does a textual conversation with the chat bot, it is sent to the IBM cloud. A response is generated and showed in user interface accordingly. We have connected the application with a cloud database, where some specific information: age, gender and employment status from the textual conversation are stored. The admins of 'Muktomon' can fetch data from database to create CSV file and store in a repository. Some analysis like, which gender and age range is facing more mental health issues and which mental health issues are affecting the end-users more, is done using the stored data, which is only accessed by data analysts and administrator.

Doctors can be registered in the application, while users may select any doctor from the list, and then the application will connect him/her with the selected doctor. A number of authentic news or information sources (e.g., WHO, UNICEF, etc.) are added in the authentic news source module of 'Muktomon'. Users can select any of the sources to further explore the news/information.

5 Evaluation of the System

The System Usability Scale (SUS) was used to evaluate the usability of the 'MuktoMon' app [21]. This section will briefly discuss the participant's profile, the study procedure, and the results of the evaluation.

5.1 Participant's Profile

To conduct the evaluation study, a total of 20 participants (12 male, 8 female) of different age groups varying from 17 to 51 years were recruited. Among them, 8 were undergraduate students, 3 of them were housewives, 7 serving at different government and non-government organizations and the remaining 2 were unemployed. All participants were familiar with the internet and smartphone. None of them have used 'Muktomon' or any other application related to mental health before.

5.2 Study Procedure

For each test-session, the following steps were followed. Firstly, participants were informed about the purpose of this study, and a participant consent form was signed. Then, the app was demonstrated and asked them to explore the app for 4–6 min. Finally, the participants were asked to provide their opinion to ten questionnaires (statements) developed following the guidelines suggested in the SUS method [21]. The questionnaires cover factors related to the necessity and usability of the application, that includes: I think that I would like to use this system frequently (Q1), I found the system unnecessarily complex (Q2), I thought the system was easy to use (Q3), I think that I would need the support of a technical person to be able to use this system (Q4), I found the various functions in this system were well integrated (Q5), I thought there was too much inconsistency in this system (Q6), I would imagine that most people would learn to use this system very quickly (Q7), I found the system very cumbersome to use (Q8), I felt very confident using the system (Q9), I needed to learn a lot of things before I could get going with this system (Q10) [21]. Participants ranked each question from 1 to 5 (1: strongly disagree and 5: strongly agree) based on how much they agree with the given statements.

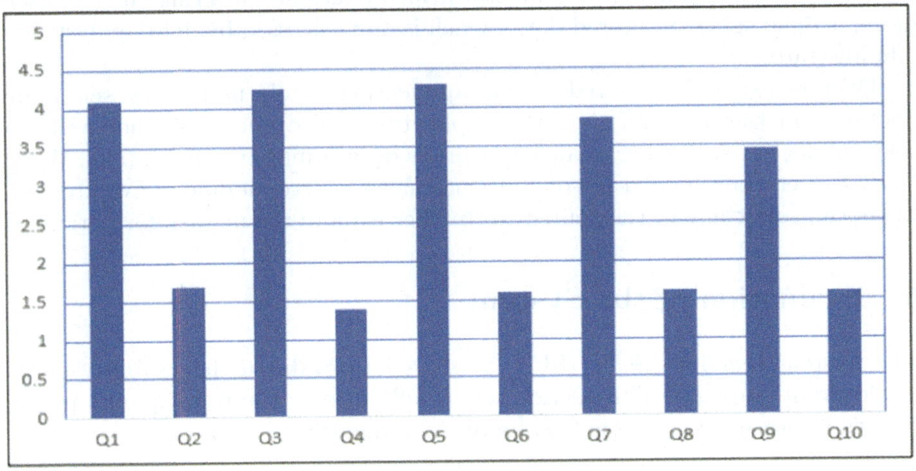

Fig. 4. Average SUS scores

Table 1. SUS scores of the studied participants

Participants	P1	P2	P3	P4	P5	P6	P7	P8	P9	P10	P11	P12	P13	P14	P15	P16	P17	P18	P19	P20
Total scores	29	34	34	31	33	32	36	34	35	35	32	35	28	31	30	26	29	29	33	33
Final scores	72.5	85	85	77.5	82.5	80	90	85	87.5	87.5	80	87.5	70	77.5	75	65	72.5	72.5	82.5	82.5

5.3 Analysis and Result

The average scores of the individual SUS questions are shown in Fig. 4. All odd-numbered questions represent positive statements and all even-numbered questions represent negative statements. Therefore, it indicates that the participants agreed (scores are above 3) with the positive statements and disagreed (scores are below 2) with the negative statements. Then the SUS scores were calculated as suggested in [21]. Firstly, for each of the odd-numbered questions, we have subtracted 1 from the score. Secondly, for each of the even-numbered questions, we have subtracted their value from 5. Thirdly, we have added up the total scores and for the final score, we have multiplied the total score by 2.5 and finally, we have calculated the average of the final scores. The results of SUS scores are presented in Table 1. The overall SUS score of the developed application is 79.875 (average of the final scores). According to [21], SUS scores 0–64 represents not acceptable; 65–84 represents acceptable and 85–100 represents excellent. Thus, the resultant SUS score indicates that the 'Muktomon' application is usable and useful to the participants.

The study results were also analyzed using inferential statistics to explore the effects of the participant's age and gender on the usability of the developed application. For this, a t-test was conducted considering the two factors of participant's profile. These are age (below the age of 25 and equal or above 25) and gender (male and female). The number of participants below the age of 25 was 10, while the remaining 10 was the age of equal or above 25. Considering the age factor, the t-value is 1.812 and the p-value is 0.067. Considering the gender factor, the t-value is 1.770 and the p-value is 0.381. Thus, the results of the t-test indicated that no statistically significant difference is observed between these groups; and the developed application is usable and useful irrespective of participant's profile in terms of their age and gender.

6 Conclusion

Mental health is a very important concern for any individual, which has become a more vital issue during the COVID-19 pandemic. Since people are not allowed to go out physically and depend more on digital interventions, so everything needs to be done virtually. In this regard, a digital solution 'Muktomon' app is developed to address several features for overcoming the mental health problems of the Bangladeshi people experienced during the COVID-19 pandemic. The application gives services for virtual therapy, AI-based chatbot, a bridge to doctors, and authentic sources of information to address the user requirements as revealed through the requirement elicitation study. A SUS based evaluation study was conducted that showed that the application is usable and useful to the end-users as well.

This research has a few limitations. Firstly, this application has not been implemented on a large scale. Secondly, the participants recruited for the evaluation study was not adequate. Finally, the usability evaluation of the developed application was evaluated by only the SUS method. The potential future research may be conducted by including few other features such as tailored feedback or recommendation from patients in the application; and adapting the machine learning technique to predict mental health problems. An extensive evaluation study following other methods involving adequate number of participants with the different backgrounds may be conducted to validate and generalize the outcomes of this research.

References

1. Laato, S., Islam, A.N., Islam, M.N., Whelan, E.: What drives unverified information sharing and cyberchondria during the COVID-19 pandemic? Eur. J. Inf. Syst. **29**(3), 288–305 (2020)
2. Ozili, P.K., Arun, T.: Spillover of COVID-19: impact on the global economy. Available at SSRN 3562570 (2020)
3. Lalon, R.M.: COVID-19 vs. Bangladesh: is it possible to recover the impending economic distress amid this pandemic? J. Econom. Bus. **3**(2), 825–836 (2020)

4. Islam, M.N., Islam, A.N.: A systematic review of the digital interventions for fighting COVID-19: the Bangladesh perspective. IEEE Access **8**, 114078–114087 (2020)
5. Anwar, S., Nasrullah, M., Hosen, M.J.: COVID-19 and Bangladesh: challenges and how to address them. Front. Public Health **8**, 154 (2020)
6. Barua, S., et al.: Understanding coronanomics: the economic implications of the coronavirus (Covid-19) pandemic. SSRN Electron. J. (2020). https://doi.org/10/ggq92n
7. Shammi, M., Bodrud-Doza, M., Islam, A.R.M.T., Rahman, M.M.: COVID-19 pandemic, socioeconomic crisis and human stress in resource-limited settings: a case from Bangladesh. Heliyon **6**, e04063 (2020)
8. Perera, W.: Bangladesh government downplays COVID-19 threat as job losses mount. Accessed 27 Mar 2020
9. Bodrud-Doza, M., Shammi, M., Bahlman, L., Islam, A.R.M., Rahman, M., et al.: Psychosocial and socio-economic crisis in Bangladesh due to COVID-19 pandemic: a perception-based assessment. Front. Public Health **8**, 341 (2020)
10. Yeasmin, S., et al.: Impact of COVID-19 pandemic on the mental health of children in Bangladesh: a cross-sectional study. Children Youth Serv. Rev. **117**, 105–277 (2020)
11. Al Banna, M.H., et al.: The impact of the COVID-19 pandemic on the mental health of the adult population in Bangladesh: a nationwide cross-sectional study, 1–12 (2020)
12. Islam, M.N., Ahmed, M.A., Islam, A.N.: Chakuri-Bazaar: a mobile application for illiterate and semi-literate people for searching employment. Int. J. Mob. Hum. Comput. Interact. (IJMHCI) **12**(2), 22–39 (2020)
13. Karim, M.M., et al.: Mobile health applications in Bangladesh: a state-of-the-art. In: 3rd International Conference on Electrical Engineering and Information Communication Technology (ICEEICT), pp. 1–5. IEEE (2016)
14. Rajkumar, R.P.: COVID-19 and mental health: a review of the existing literature. Asian J. Psychiatr. **52**, 102066 (2020)
15. Khattar, A., Jain, P.R., Quadri, S.: Effects of the disastrous pandemic COVID-19 on learning styles, activities and mental health of young Indian students-a machine learning approach. In: 4th International Conference on Intelligent Computing and Control Systems (ICICCS), pp. 1190–1195. IEEE (2020)
16. Oh, K.J., Lee, D., Ko, B., Choi, H.J.: A chatbot for psychiatric counseling in mental healthcare service based on emotional dialogue analysis and sentence generation. In: 18th IEEE International Conference on Mobile Data Management (MDM), pp. 371–375. IEEE (2017)
17. Hossain, K.A., Roy, S., Ullah, M.M., Kabir, R., Arafat, S.Y.: COVID-19 and mental health challenges in Bangladesh. Adv. Med. Dental Health Sci. **3**(2), 31–33 (2020)
18. Teles, A., et al.: Mobile mental health: a review of applications for depression assistance. In: IEEE 32nd International Symposium on Computer-Based Medical Systems (CBMS), pp. 708–713. IEEE (2019)
19. Islam, M.N., Islam, I., Munim, K.M., Islam, A.N.: A review on the mobile applications developed for COVID-19: an exploratory analysis. IEEE Access **8**, 145601–145610 (2020)
20. Thorat, S.A., Jadhav, V.: A review on implementation issues of rule-based chatbot systems. Available at SSRN 3567047 (2020)
21. McLellan, S., Muddimer, A., Peres, S.C.: The effect of experience on system usability scale ratings. J. Usability Stud. **7**(2), 56–67 (2012)
22. Islam, M.N., Inan, T.T., Islam, A.N.: COVID-19 and the rohingya refugees in Bangladesh: the challenges and recommendations. Asia Pac. J. Public Health **32**(5), 283–284 (2020)

Navigation for the Vision Impaired with Spatial Audio and Ultrasonic Obstacle Sensors

Muhd Amin Hj Fauzul and Noor Deenina Hj Mohd Salleh(✉)

School of Computing and Informatics, Universiti Teknologi Brunei, Bandar Seri Begawan, Brunei Darussalam
amin.fauzul@outlook.com, deenina.salleh@utb.edu.bn

Abstract. We develop a visual assistive system to aid the visually impaired users for safe and convenient navigation in indoor and outdoor environments. The system has two main components: a mobile app and an obstacle sensor. The mobile app makes use of the microelectromechanical sensors inside the smartphone, location services, and Google Maps to provide audio cues directions for the user. The obstacle sensor employs ultrasonic sensors to detect obstacles and provide haptic feedback. The obstacle avoidance device can be attached to a traditional probing cane, and the device vibrates the probing cane handle at different intensity according to the nature of obstacles. Spatial sound cues are generated based on the spatial distance and direction of the current location to the desired destination. We discuss the system design and report the testing to evaluate the device effectiveness for obstacle distance feedback as well as the effectiveness of the spatial sound cues for navigation feedback.

Keywords: Blindness · Spatial audio · Ultrasonic distance sensor · Vision impaired

1 Introduction

In 2019, WHO has stated there are approximately 2.2 billion people all around the world currently living with some form of vision impairment [1] which includes low vision and blindness. Vision impairment should not stop the people suffering from it to lose their independence in getting to and from places. Remembering the layout and where things are is a technique used to get around familiar places. However, travelling in unfamiliar places requires mental memory of addresses, various other senses, e.g. sound, touch, smell, and mobility skills. Devices which the vision impaired persons can use to assist their travel include probing cane, guide dogs, sighted guides, and other assistive technology devices.

Visual assistive technology consists of electronic devices which provides detection and localisation of objects for those in need of visual assistance by giving a sense of the external environment and help them determine the dimensions, range, and height of objects. Vision substitution is one of the categories of visual assistive technology, and it is the conveying of visual information via non-visual display through tactile or auditory

W. S. H. Suhaili et al. (Eds.): CIIS 2020, AISC 1321, pp. 43–53, 2021.
https://doi.org/10.1007/978-3-030-68133-3_5

information or both. This category includes devices that gather information about the surrounding environment through sensor cameras, sonar, or laser scanners [2], devices that provide pedestrians with directions in unfamiliar places [3], and devices that use technology such as GPS to determine the position of its holder. The essential factors [4] to consider when developing the technology are the cognitive demands associated with interpreting visual information non-visually, the selection of environmental information to be conveyed, and the form and function which includes the aesthetics and effective operation of the technology.

This paper is organised as follows; in Sect. 2, we review the work related to our investigation. Section 3 is where we discuss and define the design and implementation of our experiment in detail. In Sect. 4, we report the discussions of our experiment and analyse the results. Finally, we conclude our findings and discuss the future work in Sect. 5.

2 Related Works

Our experiment investigates the effectiveness of non-visual navigation and the development of visual assistive technology which utilises ultrasonic sensors for obstacle detection, uses auditory and haptic feedback to navigate, and employ *GPS* to implement navigational functions. Elmannai and Elleithy [5] discussed and compared 25 types and models of wearable and portable assistive devices for the vision impaired. We look into systems similar to our work and study the results of their experiment.

The Smart Cane [6] was equipped with a sensor system consisting of ultrasonic distance sensors and a fuzzy controller. It also came with a speaker to give messages as feedback and special vibrator gloves which was used to provide specific vibration for each finger with each vibration pattern having a different meaning. In a noisy environment, the user may not be able to hear the audio output.

Eye Substitution [7] was an embedded device used for finding directions and navigation by using two *HC-SR04 ultrasonic sensors* to overcome the issue of the narrow cone angle from using a single sensor. Finding directions involved developing an Android application to utilise the *GPS*, improved *GSM*, and *GPRS* to get the user's location and generate better directions. The user is required to hold the device, and upon detecting obstacles, the vibration motor would alert the user.

A similar work explored the development of a navigator with a 3D sound system developed using *Raspberry Pi* and GPS receiver [8]. The system calculates the distance between the user and an object using information from a gyroscope and magnetic compass. Pedestrian routes were generated using software modules and the process includes specifying the destination's which were then translated to geographic coordinates and then, obtaining the user's location via GPS to be used for route traversal. Voice prompts and audible instructions were used to navigate the user to their destination.

ELC [9] was an enhanced traditional cane which was able to provide accurate detection of objects around the user, mainly above their waistline which is considered to be one of the causes of serious injuries for the vision impaired. The detection provides tactile feedback which varies in frequency depending on the distance of the user to the object. Part of the cane was a small grip with embedded electronics consisting of an ultrasonic distance sensor, micro-motor actuator, and a 9 V battery. The cane had a minimal form factor and aesthetics but lacked the navigational capability.

An ultrasonic-based cane [10] was developed as a continuation of a C-5 laser cane to replace the laser and avoid its associated risks. The cane made use of 3 ultrasonic distance sensors to detect ground and aerial obstacles 5 cm–150 cm in range. Upon detection, a control signal would be sent wirelessly by X-bee-S1 trans-receiver module to its receiving unit mounted on the shoulder of the user along with a buzzer used to alert the user of any obstacle.

Similar work also used three ultrasonic distance sensors attached to a cane but was used to detect obstacles positioned at the front, left front, and right front of the user. Each sensor was used to collect a different range of distance. The top sensor was used for detecting overhead obstacles while the other two sensors were used for detecting front obstacles. The distance to the obstacle was calculated by acquiring separate reading of travelled times between the ultrasonic transmit and receive modules. The work has shown that the system was able to detect obstacles in four directions – front, left front, right front, and overhead.

A wearable assistive navigator system called Silicon Eyes [11] developed which was able to detect the current location of the user using GSM and GPS coordinator. The system used haptic feedback to guide, and included auditory information features which provide time and date and even, colour of objects. The device involved embedded electronics within a silicon glove which consisted of a microcontroller, colour sensor, light/temperature sensor, and SONAR. The user would choose their desired destination and they would be directed using the MEMS accelerometer and magnetometer in addition to the audio instructions via the MP3 decoder. The system also supported a touch keyboard using Braille to enter information and sending current location via SMS to specified contact in case of emergency.

Another similar development involved the use of an assistive headset consisting of four ultrasonic distance sensors with two sensors positioned on each side to detect obstacles on the left and right side of the user. The system included a recording storage for recording the recommended directions in the form of audio messages. The system played a different audio message to identify six directions but the study showed that it was not sufficient to provide precise navigation for the user for both indoor and outdoor use. Using a headset will also obscure external noises which the vision impaired relies on for mobility and orientation.

Most of the similar works explored only addressed either the mobility aspect or the navigational capability of navigation for the vision impaired. This work is focused on indoor and outdoor navigation for the vision impaired which explores adding navigation functions by experimenting with spatial auditory information in addition to adding the ability for traditional mobility tools for the vision impaired to detect obstacles in the upper body area to ensure safety during navigation.

3 System Design and Implementation

In this work, we focused on addressing both the indoor and outdoor navigation problems for the vision impaired. Integrating and combining various types of visual of assistive technologies (see Fig. 1) is an approach to utilise their strengths and cover the weaknesses of each technology. We design and develop a visual assistive technology device with embedded electronics which can be attached to any traditional long cane for detecting obstacles in the upper body area and provide timely feedback via tactile information to the user depending on the distance of the obstacle. *Raspberry Pi Zero W* was used to control the device. The controller was used to control the ultrasonic distance sensor and drive the small vibration motor. An ultrasonic distance sensor was used for measuring the distance for obstacle detection while the small vibration motor provides the tactile feedback. The Pi Zero W was chosen for this experiment for its small form factor which is suitable for our low-profile design for the obstacle detection device and allowing it to be easily mountable. In addition, its built-in Bluetooth 4.0 which enables low power consumption connectivity via the *Bluetooth Low Energy (BLE)* protocol allows our mobile app to communicate with the device and exchange information which lets us store all sensor readings on the app and on the cloud for further analysis. An enclosure for the device was developed and printed with an FDM 3D printer and it was designed to be attachable to a cane and also as a necklace.

The mobile app developed for the experiment focused on addressing outdoor navigation by helping the user to navigate to a nearby destination with the aid of 3D audio in the form of spatial sound cues which helps the user to locate and determine the position of their destination. Distance to destination was indicated by the audio playback rate. Hence, the app requires the use of stereo earbuds to clearly hear the audio cues. The app also requires location services for determining the current location of the user and their distance to the destination. The accuracy of the location services can be unreliable indoors and thus, the mobile app is only suitable for use in outdoor navigation. The user could select their destination through the embedded Google Maps. Development of the app was done using Flutter, a hybrid application development framework, which allows building the app for both Android and iOS with a single codebase. The spatial audio cues were generated using a plugin called Envelop for Live (E4L). The plugin works with Ableton Live, an audio production software, and provides immersive audio production tools for the spatialisation of spherical sound.

The device and mobile app were designed to work in tandem and integrating them provides the navigational capability to enhance the mobility of the vision impaired. Both components could be connected via Bluetooth but for this experiment, its only purpose

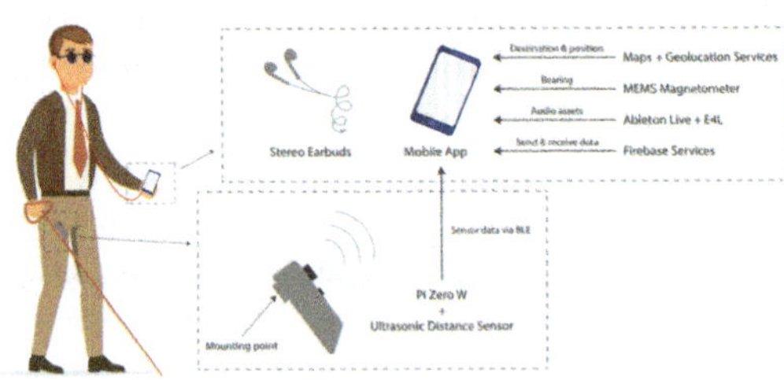

Fig. 1. The two main components of the developed visual assistive technology: mobile app for non-visual navigation using spatial audio cues, maps, geolocation services, and magnetometer and the obstacle detector device which can be mounted on a traditional long cane

was to send the sensor data to the mobile app for display and storing. The functions of each component were able to work independently (Fig. 2).

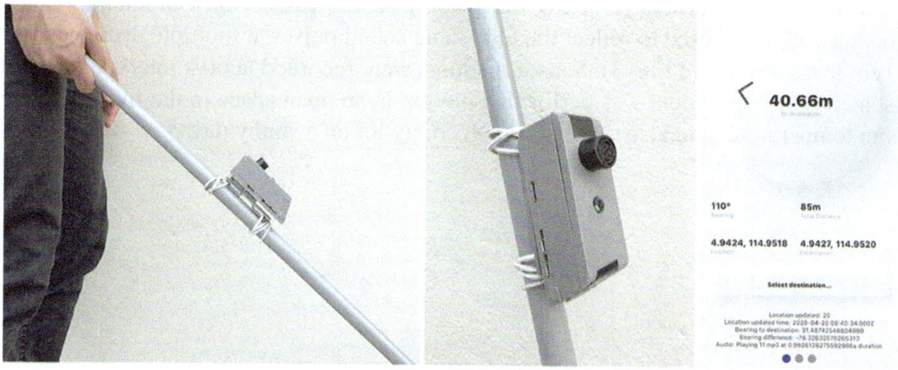

Fig. 2. The obstacle detection device prototype with a single sensor setup and vibration motor to provide haptic feedback (left) and a screenshot of the log screen from the mobile app to monitor the current status of the geolocation services, distance and bearing to destination, and the spatial audio clip being played (right)

4 Experiments and Discussions

We conduct experiments to observe the accuracy and effectiveness of the obstacle detection device by performing distance measurement tests for multiple ultrasonic distance sensors. Experiments conducted for the mobile app involves producing spatial audio using digital audio workstation tools to determine if it can be used to accurately convey positions in non-visual information.

4.1 Obstacle Detection Experiments

Experiments for obstacle detection were conducted using four ultrasonic distance sensors to compare their effective measuring distance and angle. Each sensor varies in form, maximum distance range, and measuring angle (see Table 1).

Table 1. Technical specifications for the ultrasonic distance sensors

Specification	HC-SR04	US-015	JSN-SR04T	MB1010
Operating voltage	5 V DC	5 V DC	3.3–5 V DC	2.5–5 V DC
Working current	2 mA	2.2 mA	5 mA	2 mA
Resolution	0.3 cm	0.1 cm	0.5 cm	2.5 cm
Ranging distance	2–400 cm	2–400 cm	25–450 cm	0–645 cm

The experiment was performed by controlling all the sensors with an Arduino Uno with the sensors mounted together on the same board. The sensors were then fired in sequence with a 250 ms delay added in between. This was to avoid any transmitted signal from other sensors affecting the reading of the current measurement. A large wooden board was used to reflect the ultrasonic sound pulses at multiple distances with 50 cm increments (see Fig. 3). Sensor readings were recorded at 60-s intervals for each distance. The experiment was performed indoor, in an open space in the living room at room temperature, and also outdoor, on an empty lot on a sunny day.

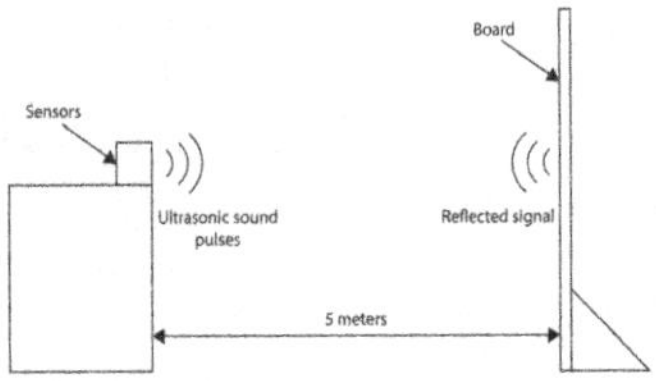

Fig. 3. Ultrasonic distance sensors distance measurement test setup

Results has shown that the HC-SR04, US-015, and JSN-SR04T could measure up to 3.5 m but measure accurately when detecting objects from 2 m. The US-015 was least accurate outdoors compared to both the HC-SR04 and JSN-SR04T. The MB1010 had consistent accuracy for both indoor and outdoor, being able to measure up to 5 m but failed to detect objects from a distance of less than 50 cm. Objects from short distances was accurately measured while using the HC-SR04 and US-015.

Measuring the effective angle for each sensor was done separately. The sensor was mounted on a small breadboard with a small arrow attached to its bottom to indicate the angle measurements on the protractor. An object measuring 217 mm in width and 107 mm in height was placed 300 mm away from the sensors. Angle measurements were recorded at 10° intervals starting from an angle of 30° to 150°. This experiment was only done indoors. As shown in Fig. 4, the JSN-SR04T and MB1010 sensors were able to measure with a wider cone of at least 10°.

We can observe the effectiveness of each sensor in different use cases and applications from both experiments. The MB1010 was eventually used in the device implementation because it was able to measure longer distances consistently and have the wide detection cone which made it efficient in detecting objects even from further away and allow feedback to be provided promptly.

4.2 Spatial Audio Experiments

The first step in producing the spatial sound cues was finding a short audio clip to use, preferably with less than a second duration e.g. beep sound. This clip would then be imported into an Ableton Live project as an audio track to allow the E4L plugins to be applied to the track as an audio effect. The effect used was the E4L Source Panner which can be configured to spatialise the sound of the audio clip into a 16-channel Ambisonics signal that will routed to the E4L Master Bus.

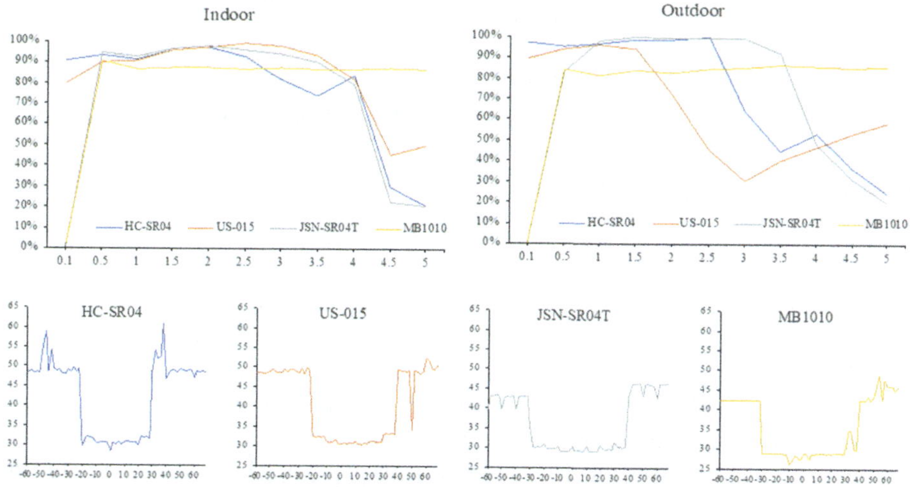

Fig. 4. Results of the indoor and outdoor distance measurement experiment (top) and the results of the angle measurement experiment (bottom). The horizontal for the top graphs shows the distance measured in meters and the vertical axis shows the accuracy of the sensor obtained from the average of the sensor measurement at 60-s intervals. The horizontal axis for the bottom graphs shows the measured angle obtained from the average of the sensor measurement at 5-s intervals and the vertical axis shows the distance measured in centimetres

15 duplicate tracks were then added to create a total of 16 audio tracks with each track configured differently. Indicating location information can be done by producing eight different sounds to distinguish each position but this experiment was set to a resolution of 16 positions to allow more accuracy in pinpointing locations. This was done by setting different azimuth value for each track to indicate the different positions (see Table 2). Each track was then exported as individual audio clips and used as assets for the mobile app.

Table 2. E4L Source Panner parameter settings

Settings	Value	Description
Range	1	Maximum distance represented by the panner
Coordinate mode	Polar	Whether coordinates are mapped in the polar or cartesian system
Mono/Stereo	Stereo	Whether the device operates in mono or stereo
Azimuth	0	The azimuth of the sound source around the X-Y plane
Elevation spread	−30.47	The elevation of the sound source
Radius	1	The radius from centre of the sound source

(continued)

Table 2. (*continued*)

Settings	Value	Description
Azimuth spread	21	Amount by which the left and right channels are spread in the angle of azimuth
Elevation spread	0	Amount by which the left and right channels are spread in the angle of elevation

The mobile app used location services to determine the latitude and longitude of the user. The user could select their destination using the embedded Google Maps and the latitude and longitude of the destination was obtained for calculating the bearing from the user's location. A MEMS magnetometer within the smartphone was utilised to determine the direction where the user was facing. The user was facing the correct direction if the reading from the magnetometer and the bearing between the user and the destination were equal. The mobile app checked the two values at a specified interval and depending on the difference between the values, a different spatial sound cue audio file would be played to indicate which position the user should be facing. The concept is presented as shown in Fig. 5 which indicates the direction the user should be facing to reach their destination which they can identify from the audio file being played. The distance between the user and destination was also updated for each location updates and this value was used to change interval duration for the audio playback. Longer interval duration meant that the user was further from the destination and shorter duration for when the user was closer.

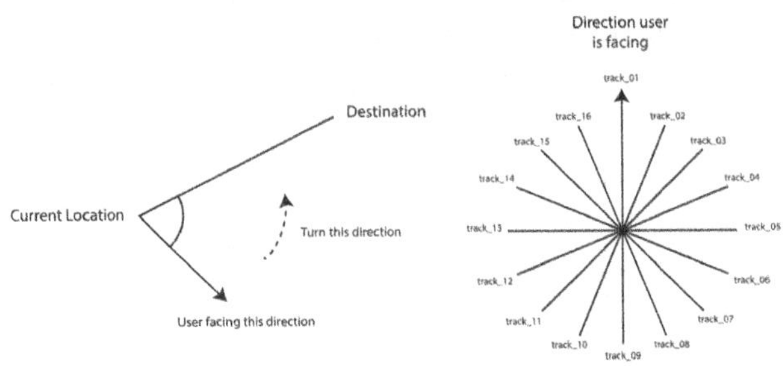

Fig. 5. A function in the mobile app uses the current location of the user using geolocation services and the magnetometer value to find the user's bearing from the destination and calculates the direction of the destination from the user's location and plays the appropriate spatial audio track to indicate the direction the user should be facing

This spatial audio experiment was conducted by distributing a short survey through Google Forms. The survey included a test where the participants were instructed to listen to eight spatial sound cues and indicate from which position the sound was coming, either from the front, front right, right, back right, back, back left, left, and front right. The participants were required to use a pair of stereo earbuds. There were ten participants consisting of people with varying visual acuity. The scores for each participant were obtained by awarding one (1) point for identifying the exact position, 0.5 point for half-right identifications, and 0 point for any wrong identifications as shown in Table 3 and Table 4. Results has shown that 67% of the participants found it difficult to distinguish between each sound and were unable to clearly identify the position each sound was coming from, especially when differentiating the 4 intermediate directions. The front and back cues sounded similar while the left and right cues were easier to differentiate. The results have also shown that there is no correlation between visual acuity and the participants ability to differentiate the sounds. A different approach to producing the spatial audio, using a different configuration, or using a mixture of sounds are some of the ways to rectify this problem. Improving the audio production and quality could lead to more distinctive sounds and may help in identifying the relationship between the proficiency of people with low vision and those with normal vision in processing spatial auditory information.

Table 3. Scores for each test participant with normal to moderate visual acuity

No	Left	Right	Front	Back	Front right	Back left	Front left	Back right	Score
1	0	0.5	1	0.5	1	0.5	0.5	0.5	0.5625
2	0	0.5	1	0.5	0.5	0.5	0.5	0.5	0.5
3	0.5	1	0.5	0	1	0.5	0.5	1	0.625
4	1	1	0.5	0	0.5	1	0.5	0.5	0.625
5	1	1	0.5	0.5	0.5	0	0	0.5	0.5
6	1	1	1	1	0.5	0.5	0.5	0.5	0.75
7	1	1	1	1	0.5	1	0.5	1	0.875
8	0	0.5	0	1	1	0.5	0.5	0.5	0.5
9	1	1	1	1	0.5	0.5	1	0.5	0.8125
10	1	1	0	0	1	0.5	0.5	1	0.625

Table 4. Scores for each test participant with low vision

No	Left	Right	Front	Back	Front right	Back left	Front left	Back right	Score
1	1	1	1	0	0	1	0.5	0.5	0.625
2	0	1	0	1	1	0.5	0.5	1	0.625
3	1	1	0	0.5	1	0.5	1	1	0.75
4	0	0	0	1	0	0	0	0	0.125
5	1	1	0	0	0.5	1	1	1	0.6875

5 Conclusions and Future Work

In this work, we demonstrated how we integrated a visual assistive technology device and a mobile app to develop a combination of visual assistive technologies to incorporate the capabilities of both components to aid the vision impaired in navigation. Results obtained from the distance and angle measurement tests suggest that ultrasonic distance sensors can be effectively used for detecting obstacles and are able to provide measurement readings with enough time to alert the user and for them to react and avoid the obstacles. Results of the spatial audio experiment have shown that it can be used to indicate location as non-visual information through auditory channels and thus, be used to facilitate navigation for people suffering from low vision. In future work, we could explore different methods and techniques of producing spatial audio, using sophisticated audio recording equipment or mixing different sounds together to create more defined and distinguishable sounds to accurately depict the directions. We could also attempt a multi-sensor setup to overcome the issue of a narrow cone angle of the sensors and making use of different sensors for handling various use cases. Extend functionalities by adding a camera to implement capabilities such as detecting colours, reading text via optical character recognition, and object detection via image processing.

Acknowledgements. The authors wish to express their very sincere thanks to the peer reviewers and Dr. Somnuk Phon-Amnuaisuk for all their helpful suggestions and valuable feedback which have improved the paper considerably.

References

1. Sensory Functions, Disability and Rehabilitation, World Health Organization. World report on vision, Switzerland, p. 26 (2019)
2. Liu, J., Liu, J.B., Xu, L.Q., Jin, W.D.: Electronic travel aids for the blind based on sensory substitution. In: Proceedings of the 5th International Conference on Computer Science & Education, Hefei, China, pp. 1328–1331 (2010)
3. Kammoun, S., Parseihian, G., Gutierrez, O., Brilhault, A., Serpa, A., Raynal, M., Oriola, B., Macé, M.-M., Auvray, M., Denis, M., Thorpe, S., Truillet, P., Katz, B., Jouffrais, C.: Navigation and space perception assistance for the visually impaired: the NAVIG project. IRBM **33**(2), 182–189 (2012)
4. Giudice, N.A, Legge, G.E.: Blind navigation and the role of technology. In: Engineering Handbook of Smart Technology for Aging, Disability, and Independence, pp. 479–500. Wiley (2008)
5. Elmannai, W., Elleithy, K.: Sensor-based assistive devices for visually-impaired people: current status, challenges, and future directions. Sensors **17**(3), 565 (2017)
6. Wahab, M.H.A., Talib, A.A., Kadir, H.A., Johari, A., Noraziah, A., Sidek, R.M., Mutalib, A.A.: Smart cane: assistive cane for visually-impaired people. Int. J. Comput. Sci. Issues (2011)
7. Bharambe, S., Thakker, R., Patil, H., Bhurchandi, K.M.: Substitute eyes for blind with navigator using android. In: Proceedings of the 2013 Texas Instruments India Educators' Conference, Bangalore, India, pp. 38–43. IEEE (2013)

8. Xiao, J., Ramdath, K., Iosilevish, M., Sigh, D., Tsakas, A.: A low cost outdoor assistive navigation system for blind people. In: Proceedings of the IEEE 8th Conference on Industrial Electronics and Applications (ICIEA), Melbourne, VIC, Australia, pp. 828–833. IEEE (2013)
9. Garcia, A.R., Fonseca, R., Duran, A.: Electronic long cane for locomotion improving on visual impaired people. A case study. In: Proceedings of the Pan American Health Care Exchanges, Rio de Janeiro, Brazil, pp. 58–61. IEEE (2011)
10. Kumar, K., Champaty, B., Uvanesh, K., Chachan, R., Pal, K., Anis, A.: Development of an ultrasonic cane as a navigation aid for the blind people. In: Proceedings of the International Conference on Control, Instrumentation, Communication and Computational Technologies (ICCICCT), Kanyakumari, India, pp. 475–479. IEEE (2014)
11. Prudhvi, B.R., Bagani, R.: Silicon eyes: GPS-GSM based navigation assistant for visually impaired using capacitive touch braille keypad and smart SMS facility. In: Proceedings of the World Congress on Computer and Information Technology (WCCIT), Sousse, Tunisia. IEEE (2013)

Smart Navigation Aids for Blind and Vision Impairment People

Safaa Najah Saud[1](✉), Lilysuriazna Raya[1], Muhammad Irsyad Abdullah[1], and Mohd Zaki Awg Isa[2]

[1] Department of Engineering and Technology, Faculty of Information Sciences and Engineering, Management and Science University, 40100 Shah Alam, Malaysia
safaaengineer@gmail.com, {lilysuriazna,irsyad}@msu.edu.my
[2] Department of Optometry and Vision Science, Faculty of Health and Life Sciences, Management and Science University, 40100 Shah Alam, Malaysia
m_zaki@msu.edu.my

Abstract. Nowadays, scientific knowledge is being developed frequently to offer versatile, safer, and sound user tendencies. At this time, a large number of prevalent techniques that have been indicated by visually impaired people are intelligent tools; however, it possesses limitations. With the recent technological innovation, it isn't easy to prolong the assistance acquire for people who have visual disabilities throughout their movability. Thus, this design provides cost-effective ultrasonic-based aids, including a hat, stick, and shoes for visually encountered people today, to achieve exclusive self-reliance and even be exempt from additional assistance. Ultrasonic sensors were able to inspect different heights, and simultaneously the user will be alerted via a buzzer. A vibrator was also implemented as a substitute gadget in minimal signal coverage areas and noisy environments. The buzzer and vibration motor are initialized when the obstacle is identified. A global positioning system (GPS) was also included in the design of aids and synchronized with google map within mounted buttons. The text message unit is also utilized by the user to send out SMS info to the saved phone numbers in the Arduino in an emergency text-request by providing the allocated longitude and altitude position. The results revealed a high accuracy of the system under different circumstances, indoor and outdoor, as feedback of different users. In conclusion, this system is considered an inexpensive, friendly user, and sensible blind assist techniques for the blind and visually impaired people.

Keywords: IoT · Blind navigation aids · Smart hat · Smart stick · Smart shoe

1 Introduction

Blindness or visual impaired condition lacks visual observation due to neurological and physiological factors [1]. For a blind pedestrian, safe mobility is one of the biggest challenges faced in their daily life. According to the World Health Organization in 2020, about 1 billion people include those with moderate or severe distance vision impairment or blindness due to unaddressed refractive errors [2], and this number is growing at

W. S. H. Suhaili et al. (Eds.): CIIS 2020, AISC 1321, pp. 54–62, 2021.
https://doi.org/10.1007/978-3-030-68133-3_6

an alarming rate. Besides, about 90% of the world's visually impaired live on a low-income budget. Without this visual functionality, blind people face many problems while doing activities in their day-to-day life as they won't identify any obstacles in their path. Therefore, there is a need for a system that can assist, guide, or navigate the blind people to carry out their day-to-day life just like ordinary people. Many types of research have been conducted to build such kind of system for blind people. The capacity to navigate and move in a specified setting without help from a third person signifies individual self-sufficiency for an individual with a visual impairment [3, 4]. This individual sovereignty is sufficiently essential to impact mental health. In addition to that, a person's inability to navigate and travel separately may lead to visual impairment. The decrease in mobility has shown that some people are depressed. It is also a matter of quality of life to travel independently.

Montello and Sas [5] navigation processes are defined as two procedures: road finding and locomotion. Wayfinding is the scheduling method for reaching a required target. During the process of finding the destination, tourists choose and schedule effective paths with significant landmarks. In planning, travelers often use aids, including physical maps, road descriptions, or electronic travel aids, such as own cognitive mapping or external assistance. Locomotion is the real-time navigation method during individual moves. During the act of navigation, most tourists use both procedures. For instance, a change in the original route may need to be planned (wayfinding) and then executed (location) when encountering and avoiding a previously unknown obstacle (locomotive) while performing a previously planned route (wayfinding).

Unfortunately, visually impaired individuals who wish to navigate independently may find it difficult to perform one or both navigational procedures [6, 7]. Effectiveness of route planning may be decreased, given that some visually impaired people often perceive the world in route rather than design. An individual may be unaware of this, and thus there may also be a locomotive impact. Independent traveling needs an individual to move effectively indoors and outdoors. Each indoor and outdoor environment has its own set of features and barriers that often require distinct abilities and strategies. Due to the distinction, orientation, and mobility tests, the evaluation of indoor and outdoor environments' capabilities assesses a person's navigation abilities [8].

Furthermore, most blind and impaired people need to depend on the traditional white cane that the visually impaired community has been used for centuries. The white cane's capacity to provide its users with independent navigation is very restricted [9, 10]. It is not easy to find barriers over the tail of a user nor detect individuals or items more than a few feet away. So far, there are many technological solutions suggested and implemented, but none improved visually impaired people's mobility and lives.

New methods have been suggested, which include guidance equipment in the white cane. Many devices use this innovation, like Guide Cane, an echolocation device that detects and leads the user from objects straight on the cane [11–13]. However, these tools and comparable devices building on white cane alter the cane's features and disturb the special navigation techniques already established with only the white cane by visually impaired individuals. A step further is the lately based Drishti system, which uses ultrasound to detect indoor objects at the customers' shoulder level. Although the scheme has shown success in finding items in an indoor environment, it only identified potential barriers and not prospective items. In order to have an inaccuracy of up to 9–14 m, the system depends exclusively on GPS data. This amount of inaccuracy is enough to remove any user in an indoor setting from a target place, such as construction. Thus, while the Drishti scheme [14] warns the user against danger and barriers beyond the waistline, it is not an efficient accessory for the visually impaired due to constraints in other fields. Scientific developments have resulted in the option to use retinal implants to restore sight in further efforts to achieve navigational independence for the visually-impaired. However, these techniques are restricted to people who retain their retinal and optic nervous functions, which exclude people who have lost vision because of diabetic retinopathy, retinal loosening, glaucoma, and other damaging conditions. In addition, the construction of these instruments needs a higher knowledge of the visual globe employing electrical signals in the retinal paths [15]. Using retinal implants remains an unlikely visual impairing alternative until this understanding is attained. Technological systems stay, therefore, the most promising candidates for the autonomy of the visually impaired. While many prior instruments include echolocation, integration with GPS, and the detection of the RFID tags [16], no system used these parts and integrated them sufficiently to make a feasible browsing acceptance for the blind society. A new device is evident since, despite considerable improvements in science and technology, the white cane remained the most commonly used and acceptable navigation instrument for the visually impaired

2 Research Design and Methodology

Malaysia's total population in 2016 is estimated at almost 38 million people, with nearly 413,000 are visually impaired, 113,000 are blind, 668,000 cataract surgeries need to be done on visually impaired eyes, and 1.3 million cataract eyes. In the blind community, the main problem is to detect any obstacles around them. From another point of view, it is also challenging for them to travel from one location to another without facing any problem. Therefore, by implementing the smart bat hat, the blind society always feels safe and independent. The hat's system vibrates and notifies the blind people to be careful when any obstacles surround them. Table 1 shows a summary of the related works, along with their advantages and disadvantages. Table 2 shows the comparison between other projects and this current project in terms of power consumption, ultrasonic sensor range, response time, and more. By looking at all this qualification, it can identify which project is better.

Figure 1a displays the invented technologies' flowchart, divided into four systems, alerting, navigation, emergency, and tracking. These systems were included in both including the indoor and outdoor systems. The indoor system operation is straightforward that consisted of the obstacle detection via any of three height-based ultrasonic sensors resulted in buzzer ON, and the vibrating motor starts operating based on the ultrasonic sensor's location detects the obstacles. Besides, the function of the outdoor system is similar to the indoor scenario. However, some additional features are added to the outdoor system, such as the mounted buttons linked to the mobile app and google navigation map and allocated for home and work addresses. Furthermore, a voice navigation system is also employed thru a Bluetooth headset to direct the user without any outside voice interruption-based.

Table 1. Summary of the related works

Author (s)	Year	Tech/detection method	Remarks	REF
Amjed S. Al-Fahoum	2013	Infrared microcontroller	• Wearable equipment • Help the blind person to navigate alone safely and to avoid any obstacles	[17]
Bousbia – Salah	2011	Navigation aid for blind people	• Able to give information to the blind about urban walking routes • Provide information on overhanging obstacles	[11]
Eklas Hossain	2011	Design of a belt for blind visually impaired people	• A belt for blind wearable around the waist • Detect stairs and holes	[18]
G. Gayathri	2014	Smart walking stick	• An intelligent walking stick that alerts visually-impaired people over obstacles	[19]
Tejal Chandekar	2017	Navigation system for visually impaired using smart shoes	• To guide a visually impaired person fluidly and to alert him/her of the obstacles • Use Android application to cater to the unique needs	[20]

Table 2. Comparison of the market provided features

System	Power consumption	Ultrasonic range	Response time	Easy of usage	Ref
Infrared microcontroller	Low	Low	Slow	Not friendly	[17]
Navigation aid for blind people	Moderate	Low	Slow	Friendly	[11]
The belt for blind visually impaired people	Moderate	Moderate	Slow	Not friendly	[18]
Smart walking stick	High	Moderate	Moderate	Friendly	[19]
Smart shoes	High	High	Moderate	Friendly	[20]

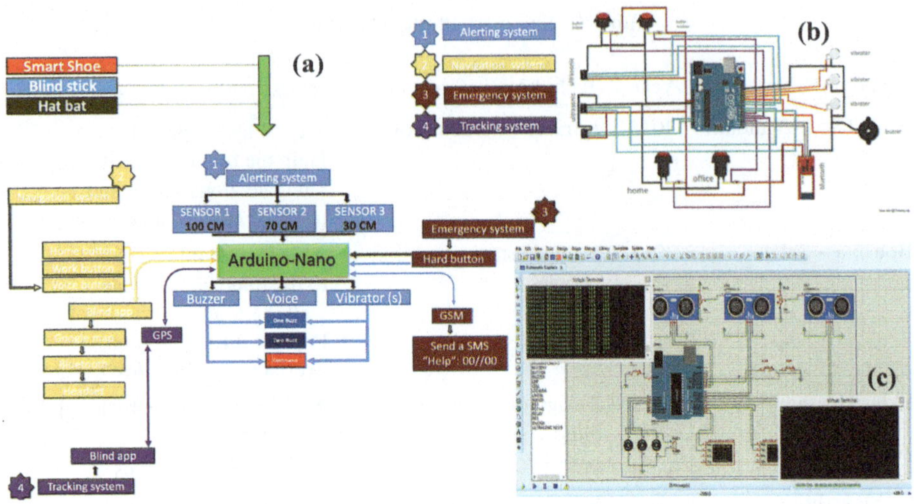

Fig. 1. (a) Flow chart, (b) Block diagram, and (c) Circuit simulation of the proposed technique

3 Results and Discussion

Designing of Schematic Diagram and Prototype: PROTEUS 8 PROFESSIONAL software was used to scheme the diagram and simulate the circuit, as shown in Figs. 1a and b. This software was implemented to ensure that the circuit connections are appropriately designed and connected.

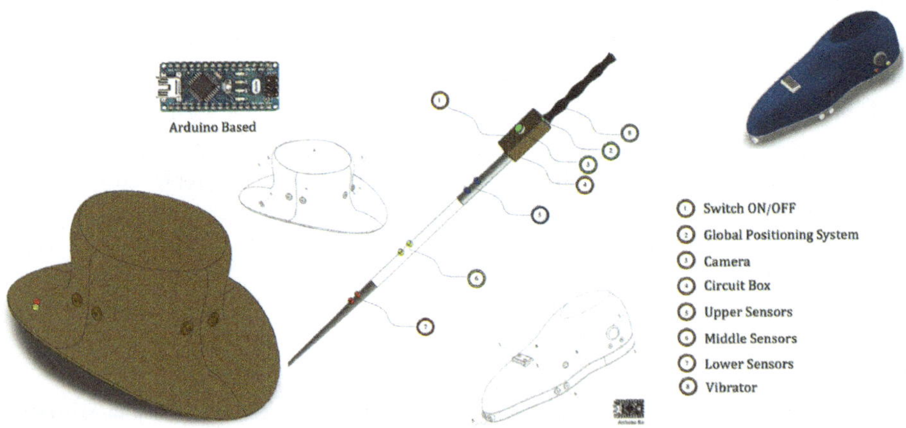

Fig. 2. Prototype design of smart: Hat, Shoe, and Stick

This circuit's main scheme consists of three ultrasonic sensors, one infrared sensor and three vibrating motors, and one buzzer. The prototype design is shown in Fig. 2, whereas designing a prototype drawing ensures how the product is in the final appearance. The prototype drawing of this project was designed by using SOLIDSWORKS18 software. In this software, the guidelines were used to design the prototype of this project.

Designing of Application: Figure 3a shows the web-application design that provides a proper and accurate location to the impaired people as being cohesive into two modes; home and office addresses. An MIT app inventor was used to design and implement this application.

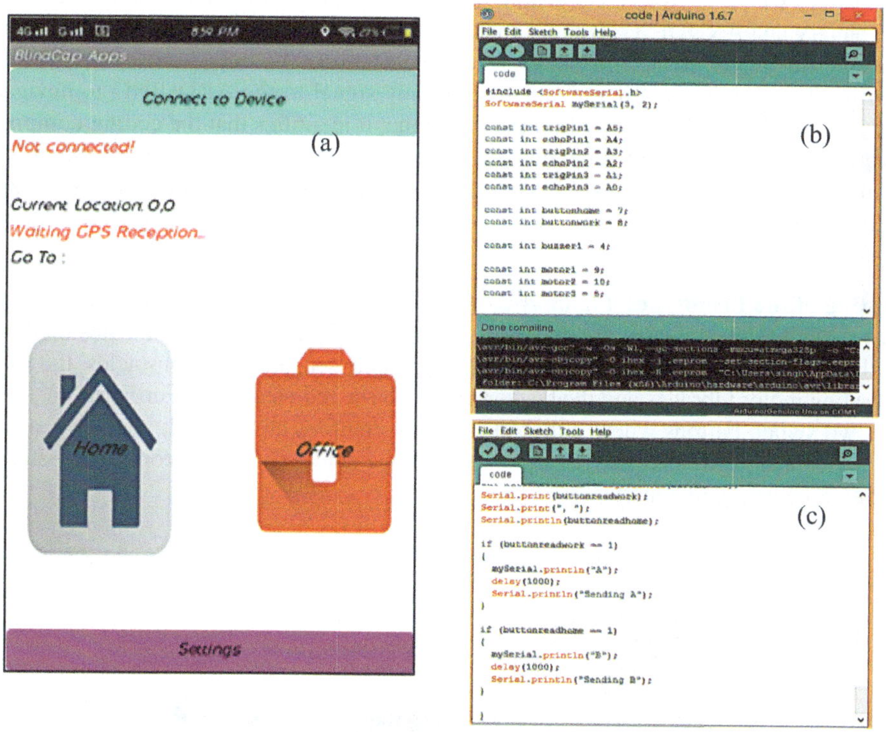

Fig. 3. (a) Design of application, (b) Indoor coding of the system, and (c) The new coding for an outdoor system.

Installing Coding in the ARDUINO UNO: Figure 3b and c illustrates the full coding of the system. The entire coding was created by using the ARDUINO software. Furthermore, the compile button was pressed in this coding file and was saved into two different files, which were in into file and hex file. The hex file was installed in the ARDUINO UNO because ARDUINO UNO can only read the hex file. The hex file was introduced by using the USB cable that was connected to the ARDUINO UNO.

Connection of Components: There were three ultrasonic sensors and vibrating motors connected to the ARDUINO UNO. A Bluetooth module connected to the system for the application to function in the system, two switch buttons connected for home and work, and a buzzer to buzz and notify the blind people. A charging module is also connected to the system to charge the battery and supply power.

Troubleshooting: During the construction of this project, several problems occurred. The first problem that occurred was about the power source; it was a 12-volt non-rechargeable battery source chosen to supply power to this project's entire system. When the troubleshooting was done, the 12-volts non-rechargeable battery drain very fast because this system requires more power source. This problem was solved by replacing the battery with a 12-volt rechargeable battery, and a charger module was connected to the battery and ARDUINO. Besides that, the other issue that occurs during this project was the outdoor system. The outdoor system, which is the home and switch button, was not functioning, and it was not sending any signal to the ARDUINO to navigate the location. When the troubleshooting was done, it identifies that the coding compiled in the ARDUINO was not accurate. After the new coding for the outdoor system was compiled into the ARDUINO, the test process was performed, whereby the system was fully functioning in send and transmitted data. Figure 3c shows the coding of the outdoor system.

Testing of the Prototype: Figure 4 demonstrates the testing process for this project. The red light indicates that the system is turned on. The yellow light indicates the level of the battery, low or high. When the front ultrasonic sensors detect obstacles, the front vibrating motor vibrates, and the buzzer sends a buzz and vice versa. Furthermore, there are two switch buttons where either of those two buttons are pressed, it sends a signal through the application to the phone, and the phone guides the blind people to its chosen location by using google map.

Fig. 4. Final prototypes linked to mobile google map using a designed mobile app

4 Conclusion

In conclusion, this system is designed especially for blind people to avoid obstacles. It is a simple, configurable, and easy to handle electronic guidance system that provides constructive assistant and support for blind people. The main functions of this system are clear path indication and environment recognition. The ultrasonic rangefinders, mini vibrating motors, and GPS navigation systems have been fully utilized so that blind people can be safe and independent. The proposed approach is a real-time system that monitors the position of the blind, which provides dual feedback to make his/her navigation more safe and secure. The only limitation that could be figured out at this stage is the limited GPS signal for the indoor system.

Acknowledgments. The author would like to thank the Management and Science University (MSU) for providing the research support under the Seed Research Grant No. SG-023-012019-FISE.

References

1. Ray, P.L., et al.: Representing vision and blindness. J. Biomed. Semant. **7**(1), 15 (2016)
2. Organization, W.H. Blindness, and vision impairment (2020). https://www.who.int/news-room/fact-sheets/detail/blindness-and-visual-impairment. Accessed Nov 2020
3. Lopresti, E.F., Bodine, C., Lewis, C.: Assistive technology for cognition [Understanding the Needs of Persons with Disabilities]. IEEE Eng. Med. Biol. Mag. **27**(2), 29–39 (2008)
4. Alkawaz, M.H., et al.: A digital eye navigators for the visually impaired. In: 2019 IEEE 9th International Conference on System Engineering and Technology (ICSET) (2019). IEEE
5. Montello, D.R., Sas, C.: Human factors of wayfinding in navigation (2006)
6. Kane, S.K., et al.: Freedom to roam: a study of mobile device adoption and accessibility for people with visual and motor disabilities. In: Proceedings of the 11th International ACM SIGACCESS Conference on Computers and Accessibility. ACM (2009)
7. Fong, S.L., et al.: Smart city bus application with QR code: a review. In: 2019 IEEE International Conference on Automatic Control and Intelligent Systems (I2CACIS). IEEE (2019)
8. Virgili, G., Rubin, G.: Orientation and mobility training for adults with low vision. Cochrane Database Syst. Rev. (5), 1–23 (2010)
9. Kane, S.K., et al.: Access overlays: improving non-visual access to large touch screens for blind users. In: Proceedings of the 24th Annual ACM Symposium on User Interface Software and Technology. ACM (2011)
10. Saud, S.N., et al.: Towards a new innovation of smart bat hat for visually impaired people. J. Manag. Sci. **16**(2), 54–66 (2018)
11. Bousbia-Salah, M., Bettayeb, M., Larbi, A.: A navigation aid for blind people. J. Intell. Robot. Syst. **64**(3–4), 387–400 (2011)
12. Kamal, M.M., et al.: Towards developing walking assistants for the visually impaired people. In: 2017 IEEE Region 10 Humanitarian Technology Conference (R10-HTC). IEEE (2017)
13. Saud, S.N., et al.: Smart walk stick for visually impaired people: design and implementation. J. Manag. Sci. **16**(2), 19–28 (2018)
14. Ran, L., Helal, S., Moore, S.: Drishti: an integrated indoor/outdoor blind navigation system and service. In: Proceedings of the Second IEEE Annual Conference on Pervasive Computing and Communications, 2004. IEEE (2004)

15. Dagnelie, G.: Psychophysical evaluation for visual prosthesis. Annu. Rev. Biomed. Eng. **10**, 339–368 (2008)
16. Yelamarthi, K., et al.: RFID and GPS integrated navigation system for the visually impaired. In: 2010 53rd IEEE International Midwest Symposium on Circuits and Systems. IEEE (2010)
17. Al-Fahoum, A.S., Al-Hmoud, H.B., Al-Fraihat, A.A.: A smart infrared microcontroller-based blind guidance system. Act. Passiv. Electron. Compon. **2013**, 1–7 (2013)
18. Hossain, E., Khan, R., Ali, A.: Design and data analysis for a belt-for-blind for visual impaired people. Int. J. Adv. Mechatron. Syst. **3**(5–6), 384–397 (2011)
19. Gayathri, G., et al.: Smart walking stick for visually impaired. IJECS **3**(3), 4057–4061 (2014)
20. Chandekar, T., et al.: Obstacle detection and navigation system for visually impaired using smart shoes. Int. Res. J. Eng. Technol. (IRJET) **4**(2), 574–577 (2017)

SMAC (Social, Mobile, Analytics, Cloud)-Based Learning Intervention for Introductory Programming – The Trend in the Past 15 Years

Salihin Rahim, Saiful Omar(✉), Thien Wan Au, and Irwan Mashadi Mashud

School of Computing and Informatics, Universiti Teknologi Brunei, Gadong 1410, Brunei
P20190006@student.utb.edu.bn, {saiful.omar,twan.au, irwan.mashud}@utb.edu.bn

Abstract. The lack of problem-solving skills among computer programming students is arguably an issue that concerns many educational sectors around the globe. Multiple innovative interventions have been proposed to mitigate this issue. This study aims to investigate the trends of using SMAC concept (social, mobile, analytics, cloud) in introductory programming from the years 2004 to 2019. A systematic literature review (SLR) was employed to explore how the existing learning interventions used SMAC elements for aiding introductory programming learning and determine their effectiveness. Fifty-seven relevant papers were included in the SLR, and the results showed that the majority of the existing interventions were related to the analytics element. The findings indicated that existing studies partly utilised the elements of SMAC for enhancing the collaboration among the students (social element), enabling ubiquitous learning (mobile element), analysing students' behaviour and predicting their performance (analytics element), and enabling coding on the cloud (cloud element). Most importantly, the SLR results presented a research gap in the use of all four SMAC elements to design a learning intervention that aids introductory programming learning.

Keywords: Introductory programming · SMAC (Social Mobile Analytics Cloud) · E-learning

1 Introduction

Learning computer programming has been considered difficult among computer science students, including those in the introductory programming course [1]. Even though several studies have reported evidence of students struggling in learning this course [2–4], there is no consensus on the main issue faced by the students [5]. However, the inability to think algorithmically due to lack of problem-solving skill was argued as one of the common issues that introductory programming students faced [6–11].

Multiple studies by [12] have confirmed the relation between the lack of problem-solving skills with students' difficulties in learning computer programming. Introductory

W. S. H. Suhaili et al. (Eds.): CIIS 2020, AISC 1321, pp. 63–74, 2021.
https://doi.org/10.1007/978-3-030-68133-3_7

computer students are often observed struggling in structuring the steps to develop a solution despite having an understanding of a specific programming problem [6]. Such difficulty is a significant concern as the ability to think algorithmically and possessing computational thinking are two essential skills that both students and employees of today's generation must possess [13]. In addition to this, [14] and [15] proved the importance of having problem-solving skills in ensuring students' positive learning performance in introductory programming courses.

Albeit being discussed in a plethora of studies, the lack of problem-solving skills among the students is still an unresolved issue that affects many educational sectors, suggesting the immaturity of this field of study. It is, therefore, necessary to deal with this problem because of several potential unintended effects, including the loss of learning motivation among the students that could eventually lead to failing the course. A study [16] estimated the global pass rate of around 67% for the introductory programming course, and several other studies also reported almost similar results with a failure rate ranging from 30 to 50% [17–19].

The issues mentioned above pose a question as to how the introductory programming students can acquire the requisite problem-solving skills. A study [12] highlighted the ineffectiveness of the conventional teaching strategies that focus more on providing students with programming *vocabulary* (syntaxes and semantics) rather than developing the students' problem-solving skills. As a result, the time spent on developing this essential skill is very limited. It is thence necessary to innovate the teaching and learning strategies for the introductory programming courses to retain students' engagement [20], and supporting the fourth goal of the Sustainable Development Goals, i.e., a quality education [21]. Many studies had shown innovation by introducing novel learning interventions to mitigate these issues. Still, there is currently no single solution that could cater to all problems faced in the learning of introductory programming.

This study aims to investigate the use of SMAC elements to alleviate the difficulty of learning introductory computer programming for the past 15 years using a systematic literature review (SLR). Drawn from the results, we discovered that learning intervention that adopts full SMAC concept to the introductory programming learning environment is lacking. We thence aim to adopt the SMAC concept into the computer programming learning environment by designing a SMAC-based learning intervention. The motivation comes from the idea of adopting a collaborative and ubiquitous learning into this researched educational domain, which is possible by integrating mobile platforms (such laptop computer, tablet, and smartphones) into its environment. Besides, the utilisation of cloud technology will help to reduce the needs for any prerequisite installation of desktop-based code editor that is often regarded as cumbersome [22]. Every student interaction with the cloud-based, collaborative, and ubiquitous learning environment could serve as the data source required for enabling further analytical process, which is crucial as it has been argued that applying analytical processes on educational data helps in enhancing education quality [23]. Therefore, we believe that SMAC concept fits this research domain due to the four disruptive technologies it possesses to achieve this new learning environment for computer programming.

The next section discussed the methodology used for the investigation. Section 3 showed the results and Sect. 4 provided a thorough discussion of the results. Section 5 concluded the paper with description of future work.

2 Methodology

This study employed SLR methodology to gather evidence on how the elements of SMAC were utilised in the existing learning interventions to aid the teaching and learning of introductory programming. The SLR employed in this study followed the guidelines as suggested by [24] to address three research questions as follows:

[RQ1] Which element is the most commonly used in aiding the introductory programming learning?
[RQ2] What are the types of interventions under each element of SMAC that have been proposed or introduced?
[RQ3] Do the collective studies suggest that the interventions of each element are effective in the learning of introductory programming?

Selection of Relevant Papers. The process of identifying papers involved developing a search strategy in a form of search string, which was developed by combining two keywords relevant to the research questions, *"Introductory Programming"* and *"Social Mobile Analytics Cloud"*, by using two Boolean operators (*"AND"* and *"OR*) as follows:

(("social mobile analytics cloud" OR ("social" AND "mobile*" AND "analytics*" AND "cloud*")) AND ("introductory programming" OR "cs1" OR "novice programming"))*

This study used electronic (automatic) and manual searches of digital resources, where four databases were utilised in its automatic searches including IEEE Xplore Digital Library, ACM Digital Library, Springer Link and ScienceDirect. Contrarily, six conference proceedings considered relevant were used to manually acquire relevant papers, including International Conference on Advanced Learning Technologies (ICALT), International Conference on Computer Science and Education (ICCSE), IEEE Global Engineering Education Conference (EDUCON), Frontiers in Education Conference (FIE), IEEE Conference on e-Learning, e-Management and e-Services (IC3e), and ACM Special Interest Group on Computer Science Education (SIGCSE).

Paper Selection Process. The selection of the papers was done in two phases. The first phase involved selecting potential papers by reading their title and abstract only. Meanwhile, the second phase involved reading the entire contents of all potential papers included after the first-phase screening and applying the inclusion and exclusion criteria as follows:

1. *Inclusion Criteria*

a. Population is students learning introductory programming from college to undergraduate education.

b. Focus is on the adoption of at least one of SMAC elements into the learning of introductory programming.
c. Results from the SMAC elements adoption in terms of its effectiveness in fostering students' learning performance must be reported.
d. Must be published between 2004 until 2019.

2. Exclusion Criteria

a. Population is not students learning introductory programming.
b. Focus is not on adopting SMAC elements into introductory programming learning.
c. No results provided on the effectiveness of SMAC adoption on students' learning performance.
d. Readability is limited to abstract only (no full text available).
e. Not written in English language.
f. Published before 2004 and after 2019.

Extraction of Relevant Data. This phase involved extracting all relevant data from each included paper based on their usefulness in addressing the SLR research questions. In general, the data extracted from each included paper consisted of abstracts, year of publication, aims and objectives, participants and sample size, research methodology, interventions used, comparator used, and research outcomes.

In facilitating the data extraction process, each included paper was classified based on the SMAC element it utilised. For instance, a paper that used social media was classified as a *"social"* paper. The extracted data and the classification labels were stored in a spreadsheet.

3 Results

3.1 Search Result

A total of 889 papers were gathered during the initial search of this SLR, but only 57 papers were considered relevant according to the inclusion criteria. In arriving this figure, a total of 133 papers were analysed by reading their full contents. The remaining 75 included papers were found irrelevant because they satisfied at least two of the exclusion criteria including the absence of proof on the effectiveness of the interventions in question, and the use of the wrong population in the study.

Additionally, the included 57 included papers were published from the years 2004 until 2015, the majority of which were from the year 2017 (11 papers) and the least from the year 2008 (one paper).

3.2 Trends in the Past 15 Years

SMAC Elements Utilisation. The results demonstrated that the analytics element was the most commonly used SMAC element after it was utilised in 29 of the included papers. Meanwhile, 28 papers used social element, 11 papers used the cloud element, and 6 papers utilised the mobile element, as shown in Table 1.

Moreover, the SLR also revealed that a single learning intervention could be structured or designed by using at least 1 SMAC element and at most three SMAC elements. The results showed that most studies (41 papers) only utilised 1 SMAC element in their learning interventions, whereas 15 papers combined 2 SMAC elements and 1 paper combined 3 SMAC elements. For brevity, it was observed that 5 papers combined social and mobile elements (S-M), 5 papers combined social and analytics elements (S-A), 5 papers combined social and cloud elements (S-C) and 2 papers combined analytics and cloud elements (A-C) in their learning interventions. Interestingly, the results also revealed that none of the included studies combines the entire 4 SMAC elements (S-M-A-C).

Table 1. Number of times each SMAC element being utilised in the included papers

Element	Paper count	References (see Appendix A)
Social	28	P1, P2, P6, P7, P8, P9, P11, P12, P13, P14, P18, P20, P21, P22, P23, P24, P26, P28, P31, P40, P41, P42, P44, P50, P51, P53, P54, P57
Mobile	6	P6, P12, P21, P26, P30, P40
Analytics	29	P3, P4, P5, P9, P10, P15, P16, P19, P25, P27, P28, P29, P32, P34, P35, P36, P37, P38, P39, P43, P44, P45, P48, P49, P51, P52, P54, P55, P56
Cloud	11	P1, P4, P13, P17, P24, P29, P33, P46, P47, P53, P54

Identified Learning Interventions. This SLR identified a total of 19 different learning interventions, where each identified learning intervention could associate to at most three elements of SMAC simultaneously. For instance, a learning intervention that incorporated social media was associated with both social and mobile elements. Additionally, some learning intervention was also found to associate with only one SMAC element, in which we referred to as a "*unique*" learning intervention type in this study.

As illustrated in Fig. 1, the social element associated with 7 interventions, of which 2 of them were unique. Moreover, the mobile element associated with 3 interventions and only 1 of them was unique. The analytics element recorded the highest number of associations, with a total of 9 interventions, and 8 of them were unique. Meanwhile, the cloud element associated with 5 interventions, with 3 of them were unique.

Equally important, this study sought to identify learning interventions with evidence of being frequently studied (referred to as "*usage frequency*" in this study) and having evidence of positive effectiveness. The results identified 7 learning interventions that possess the two evidence aforementioned, as outlined in Table 2. These 7 learning interventions recorded a usage frequency of 10%, whereas the remaining 12 of the 19 identified learning interventions recorded usage frequency of less than 7%.

Pair programming was the most commonly used learning intervention with 24.14% usage frequency (14 papers), and *cloud coding* came second with 15.52% usage frequency (9 papers) usage. *Early prediction modelling* and *group collaboration* were

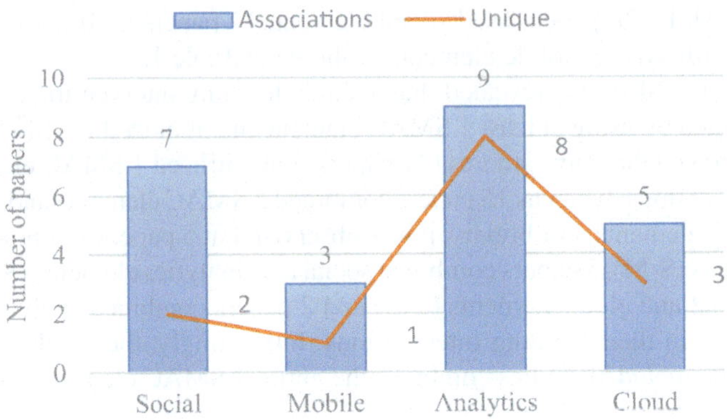

Fig. 1. Identified associated and unique learning interventions of each SMAC element

the third most commonly used learning interventions with 13.79% usage frequency (8 papers each). The fourth most frequently used interventions were the *analysis of students' programming behaviour*, *social media*, and *mobile learning*. Each of these three interventions recorded 12.07% usage frequency (7 papers each). Meanwhile, the remaining 13 interventions (not shown in Table 2) recorded a usage frequency ranging from 1.72% (1 paper) to 6.90% usage frequency (4 papers).

Table 2. Seven frequently studied learning interventions with positive effectiveness

Element	Social	Mobile	Analytics	Cloud
Interventions	Pair Programming [N: 14]	Mobile Learning [N: 7]	Early Prediction Modelling [N: 8]	Cloud Coding [N: 10]
	Group Collaboration [N: 8]	Social Media [N: 7]	Analysis of Students' Programming Behaviour [N: 7]	

Interventions Effectiveness. This study was unable to perform detailed statistical comparison, such as ANOVA, on the intervention effectives due to the heterogeneity of the included papers. However, the SLR revealed 50 (87.72%) of the 57 included papers to report positive outcomes. In comparison, 6 papers (10.53%) reported mixed outcomes and only 1 paper (1.75%) reported a negative outcome on students' learning performance.

When concentrating on the seven interventions above-mentioned (see Table 2), 10 of the 14 papers that examined pair programming reported positive outcomes with only 1 paper reported negative outcomes. Meanwhile, 7 of 8 papers that studied the effectiveness of group collaboration reported positive outcomes.

On the other hand, all 15 papers that experimented on the use of early prediction modelling and the analysis of students' programming behaviour reported positive outcomes.

Furthermore, all 7 papers experimenting the utilisation of social media for learning programming reported positive outcomes and similarly positive results were also observed in all 7 papers that studied the adoption of mobile learning to programming learning.

Equally important, 8 of the 9 papers experimenting cloud coding reported positive outcomes, while only 1 paper reported mixed outcomes (Fig. 2).

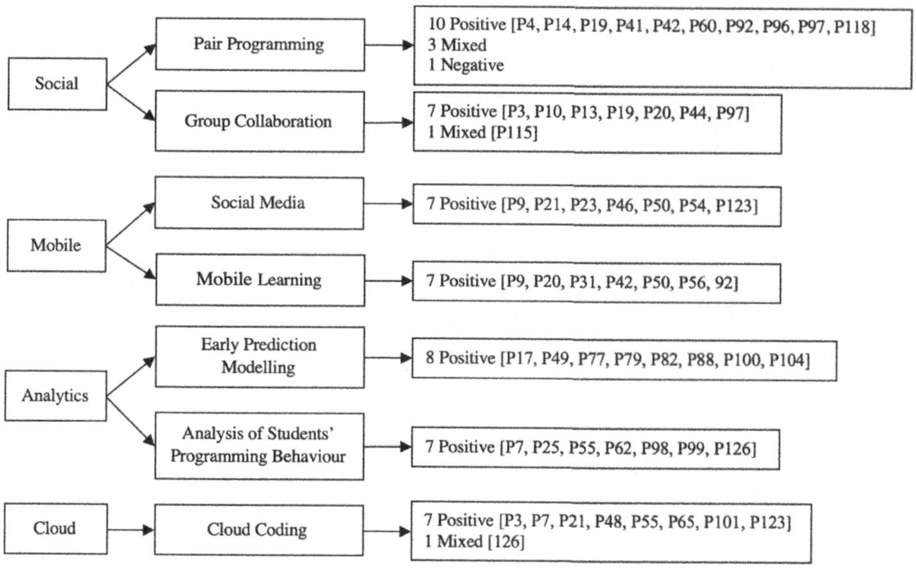

Fig. 2. Reported effectiveness of the seven frequently-used learning interventions

4 Discussions

The results have identified 19 SMAC-related learning interventions used to support the learning of the introductory programming and this answered the [RQ1] of this study. From these 19 interventions, 7 of them have been widely experimented that help to provide evidence of their effectiveness. Of these 7 interventions, the majority of the studies reported positive outcomes with only 1 negative outcome was observed (on pair programming). This finding helped to answer the [RQ2] of this study, where the results of the collective studies provided evidence that the identified interventions were effective in aiding the introductory programming learning.

Interestingly, the majority of the included papers were associated with the analytics element, which demonstrates a rising trend in adding "intelligence" into the learning environment by applying analytical techniques. This finding also answered the [RQ3] of this study. Of the 8 unique analytical techniques identified in this SLR, more interest was observed in making early predictions on students' success in the course and analysing the behaviour of students while doing programming exercises. These findings do make sense as educational sectors, specifically, have started to see a huge potential in utilising analytical techniques to improve students' learning process [25].

This study also provides evidence of more interest in improving collaboration among students by using varieties of collaborative practices other than social media, which demonstrated the importance of enhancing collaboration between students, particularly as students with collaboration skills are proven more successful and learn faster [26]. The two most frequently used collaboration practices found in this SLR were pair programming and group collaboration. The SLR also helps to deduce two observations: firstly, it is evident that there is still a limited exploration of using social media, such as *Facebook*, to aid the learning of introductory programming. Secondly, existing studies experimented more on increasing students' in-classroom collaboration rather than enhancing and promoting the engagement of remote collaboration.

Furthermore, existing studies were interested in exploring more online code editor to reduce the needs to install any code editor on students' machines. This is also accompanied by using tools to allow students' codes to be automatically assessed by an online mechanism and providing instant feedback through the online code editor.

Also, the mobile element was minimally utilised in the existing studies in comparison with other elements, which presents more opportunity for future studies to use this element to aid the learning of introductory programming courses. Another interesting observation was that all papers reviewed in this SLR used the mobile element in either of two ways: (1) as a platform for collaborating with others, or (2) as a targeted system environment of which a program is expected to be deployed, such as Android application development. This observation revealed a lacking in utilisation of any introductory programming learning-oriented mobile applications to aid the learning of this course. None of the included papers propose and experiment on this kind of mobile application and this indicates another research gap.

Most importantly, the SLR results presented a significant research gap in the use of all four SMAC elements to design a learning intervention that aids introductory programming learning. However, one issue is to determine what features that could structure an effective SMAC-based learning intervention. One possible way is by considering the *"popularity"* of the interventions identified in the SLR, which can be obtained based on their usage frequency.

5 Conclusions and Future Work

This study employed an SLR to review how existing studies in the past decade utilised the elements of SMAC concept to design their learning interventions for the teaching and learning of introductory programming. Innovations in the learning strategies for this academic field is essential, especially when the inability of students to think algorithmically due to lack of problem-solving skills is arguably one of the issues faced by the introductory programming students. Such innovation is also crucial to improve the education quality, which is one of the goals in the Sustainable Development Goals (SGDs).

The SLR conducted in this study was the first of its kind to explore the adoption of SMAC elements concept into the researched domain, where it used four electronic databases and six conference proceedings to gather relevant papers. From reviewing 133 potential papers, only 57 papers were included after applying the inclusion and exclusion criteria. For answering [RQ1], the SLR results showed an imbalance usage of the four SMAC elements, where the analytics element has a higher usage frequency compared to the other three elements (social, mobile, cloud). The cloud and mobile elements were the least frequently used elements, while the social element was found as the second most frequently used element. For answering [RQ2], this SLR identified 19 unique learning interventions associated with the SMAC elements from the existing studies. However, only seven of them have evidence of high usage frequency and positive effectiveness. In short, the SMAC elements were used for enhancing and motivating collaboration among students (social element), enabling ubiquitous learning (mobile element), analysing students' behaviour and predicting their performance (analytics element), and enabling coding on the cloud (cloud element). For answering [RQ3], the SLR results found out that 50 of the 57 papers (87.72%) reported positive outcomes that demonstrate the ability for the interventions to be replicated in future studies. Most importantly, the results revealed a significant research gap when none of the included studies combined all four SMAC elements in their learning intervention. The gap thence presents an opportunity to adapt the SMAC concept into designing a learning intervention that uses all four SMAC elements. Therefore, we aim to pursue this study in the future by developing a conceptual model and designing a new SMAC-based learning intervention based on the findings derived from the SLR. We believe that by embedding these four elements into the introductory programming learning environment, the enhancement of the students' problem-solving and collaboration skills for better learning outcomes could be achieved.

Appendix

Papers Included in the SLR

Only references to 15 included papers are presented due to page constraint. Complete references to 57 papers are available upon request.

[P1] L. C. Authors, S. Chang, C. Chen, and E. Yang, "An e-Learning system for programming languages with semi-automatic grading,". (2017).

[P2] D. McKinney and L. F. Denton, "Affective Assessment of Team Skills in Agile CS1 Labs: The Good, the Bad, and the Ugly," in *Proceedings of the 36th SIGCSE Technical Symposium on Computer Science Education*, pp. 465–469. (2005).

[P3] H. Tamada, A. Ogino, and H. Ueda, "A Framework for Programming Process Measurement and Compiling Error Interpretation for Novice Programmers," in *2011 Joint Conference of the 21st International Workshop on Software Measurement and the 6th International Conference on Software Process and Product Measurement*, pp. 233–238. (2011).

[P4] N. Ishihara and N. Funabiki, "A Proposal of Statement Fill-in-Blank Problem in Java Programming Learning Assistant System," in *2015 IIAI 4th International Congress on Advanced Applied Informatics*, pp. 247–252. (2015).

[P5] F. H. Leong, "Automatic detection of frustration of novice programmers from contextual and keystroke logs," in *2015 10th International Conference on Computer Science & Education (ICCSE)*, pp. 373–377. (2015).

[P6] M. Maleko, M. Hamilton, and D. D'Souza, "Access to mobile learning for novice programmers via social networking sites," in *2012 7th International Conference on Computer Science & Education (ICCSE)*, pp. 1533–1538. (2012).

[P7] R. Shadiev *et al.*, "Applying Unidirectional versus Reciprocal Teaching Strategies in Web-Based Environment and Their Effects on Computer Programming Learning," in *2013 IEEE 13th International Conference on Advanced Learning Technologies*, pp. 5–9. (2013).

[P8] M. Amaratunga, G. Wickramasinghe, M. Deepal, O. Perera, D. De Silva, and S. Rajapakse, "An Interactive Programming Assistance tool (iPAT) for instructors and novice programmers," in *2013 8th International Conference on Computer Science & Education*, pp. 680–684. (2013).

[P9] D. D. Phuong, F. Harada, and H. Shimakawa, "Collaborative Learning Environment to Improve Novice Programmers with Convincing Opinions in Computer Room," in *2009 International Conference on Intelligent Networking and Collaborative Systems*, pp. 61–66. (2009).

[P10] F. D. Pereira, E. H. T. Oliveira, D. Fernandes, and A. Cristea, "Early Performance Prediction for CS1 Course Students using a Combination of Machine Learning and an Evolutionary Algorithm," in *2019 IEEE 19th International Conference on Advanced Learning Technologies (ICALT)*, vol. 2161–377X, pp. 183–184. (2019).

[P11] B. Estácio *et al.*, "Evaluating Collaborative Practices in Acquiring Programming Skills: Findings of a Controlled Experiment," in *2015 29th Brazilian Symposium on Software Engineering*, pp. 150–159. (2015).

[P12] A. P. L. Ambrósio and F. M. Costa, "Evaluating the impact of PBL and tablet PCs in an algorithms and computer programming course," in *SIGCSE'10 - Proceedings of the 41st ACM Technical Symposium on Computer Science Education*, pp. 495–499. (2010).

[P13] T. Y. Sim, "Exploration on the impact of online supported methods for novice programmers," in *2015 IEEE Conference on e-Learning, e-Management and e-Services (IC3e)*, pp. 158–162. (2015).

[P14] M. Maleko, D. Nandi, M. Hamilton, D. D'Souza, and J. Harland, "Facebook Versus Blackboard for Supporting the Learning of Programming in a Fully Online Course: The Changing Face of Computing Education," in *Proceedings of the 2013 Learning and Teaching in Computing and Engineering*, pp. 83–89. (2013).

[P15] Y. Lu, I. Hsiao, and Q. Li, "Exploring Online Programming-Related Information Seeking Behaviors via Discussion Forums," in *2016 IEEE 16th International Conference on Advanced Learning Technologies (ICALT)*, pp. 283–287. (2016).

References

1. Derus, M.S.R., Ali, M.A.Z.: Difficulties in learning programming: Views of students. In: 1st Int. Conf. Curr. Issues Educ. (ICCIE 2012), pp. 74–79 (2012), September 2012
2. Govender, I., Grayson, D.J.: Pre-service and in-service teachers' experiences of learning to program in an object-oriented language. Comput. Educ. **51**(2), 874–885 (2008)

3. Kordaki, M.: A drawing and multi-representational computer environment for beginners' learning of programming using C: Design and pilot formative evaluation. Comput. Educ. **54**(1), 69–87 (2010)
4. Yang, T.C., Yang, S.J.H., Hwang, G.J.: Development of an interactive test system for students' improving learning outcomes in a computer programming course. In: Proc. - IEEE 14th Int. Conf. Adv. Learn. Technol. ICALT 2014, pp. 637–639 (2014)
5. McCall, D., Kölling, M.: Meaningful categorisation of novice programmer errors. In: 2014 IEEE Frontiers in Education Conference (FIE) Proceedings, pp. 1–8 (2014)
6. Amaratunga, M., Wickramasinghe, G., Deepal, M., Perera, O., De Silva, D., Rajapakse, S.: An Interactive Programming Assistance tool (iPAT) for instructors and novice programmers. In: 2013 8th International Conference on Computer Science & Education, pp. 680–684 (2013)
7. Ambrósio, A.P.L., Costa, F.M.: Evaluating the Impact of PBL and Tablet PCs in an algorithms and computer programming course. In: Proceedings of the 41st ACM Technical Symposium on Computer Science Education, pp. 495–499 (2010)
8. Chung, I., Hsu, C., Li, D.: A programming learning diagnostic system using case-based reasoning method. In: 2016 International Conference on System Science and Engineering (ICSSE), pp. 1–4 (2016)
9. Diez, D., Diaz, P., Aedo, I., Fernandez, C.: DEI-CHECK. Automating the assessment process to improve the informative feedback. In: 2008 38th Annual Frontiers in Education Conference, pp. F1D-18-F1D-23 (2008)
10. Intisar, C.M., Watanobe, Y.: Classification of online judge programmers based on rule extraction from self Organizing Feature Map. In: 2018 9th International Conference on Awareness Science and Technology (iCAST), pp. 313–318 (2018).
11. Sun, H., Li, B., Jiao, M.: YOJ: An online judge system designed for programming courses. In: 2014 9th International Conference on Computer Science & Education, pp. 812–816 (2014)
12. Gomes, A., Mendes, A.: A teacher's view about introductory programming teaching and learning: Difficulties, strategies and motivations. In: 2014 IEEE Frontiers in Education Conference (FIE) Proceedings, pp. 1–8 (2014)
13. Chatzinikolakis, G., Papadakis, S.: Motivating K-12 students learning fundamental Computer Science concepts with App Inventor Mobile application development in secondary education. In: International Conference on Interactive Mobile Communication Technologies and Learning, no. Imcl, pp. 152–159 (2014)
14. Alshaye, I., Tasir, Z., Jumaat, N.F.: The conceptual framework of online problem- based learning towards problem-solving ability and programming skills. In: 2019 IEEE Conference e-Learning, e-Management Eser., November 2019
15. Pillay, N., Jugoo, V.R.: An investigation into student characteristics affecting novice programming performance An Investigation into Student Characteristics Affecting Novice Programming Performance. In: ACM SIGCSE Bulletin, December 2005
16. Bennedsen, J., Caspersen, M.E.: Failure rates in introductory programming. In: ACM SIGCSE Bulletin, June 2007
17. Sim, T.Y.: Online supported learning and threshold concepts in novice programming. In: 2017 IEEE Conference on e-Learning, e-Management and e-Services (IC3e), pp. 85–90 (2017)
18. de Kereki, I.F.: Work in progress — transversal competencies contributions to Computer Science 1 course. In: 2011 Frontiers in Education Conference (FIE), pp. S3G-1–S3G-3 (2011)
19. Watson, C., Li, F.W.B., Godwin, J.L.: Predicting performance in an introductory programming course by logging and analyzing student programming behavior. In: 2013 IEEE 13th International Conference on Advanced Learning Technologies, pp. 319–323 (2013)
20. Leong, F.H.: Automatic detection of frustration of novice programmers from contextual and keystroke logs. In: 2015 10th International Conference on Computer Science & Education (ICCSE), pp. 373–377 (2015)

21. Morton, S., Pencheon, D., Squires, N.: Sustainable Development Goals (SDGs), and their implementation. Br. Med. Bull. **124**(1), 81–90 (2017)
22. Kurniawan, A., Soesanto, C., Wijaya, J.E.C.: CodeR: real-time code editor application for collaborative programming. Procedia Comput. Sci **59**, 510–519 (2015)
23. Karthick Anand Babu, A.B., Maghesh Kumar, D.: CSI Communications 1, vol. 38, p. 10, October 2014
24. Kitchenham, B., Charters, S.: Guidelines for performing systematic literature reviews in software engineering (2007)
25. Vanthienen, J., De Witte, K.: Data analytics applications in education. Data Anal. Appl. Educ. pp. 1–266, October 2017
26. Pelet, J.E., Pratt, M.A., Fauvy, S., Sinclair, J., Kalvala, S.: Learning technology for education in cloud. Commun. Comput. Inf. Sci. **533**, 43–53 (2015)

Blockchain for the Authentication and Immutability of Academic Credentials Issued in Brunei Darussalam

Zara Laila Abdul Hadi(✉) and Thien Wan Au

Universiti Teknologi Brunei, Gadong, Brunei Darussalam
zaralaila@icloud.com, twan.au@utb.edu.bn

Abstract. Blockchain technology is a distributed ledger based on a peer-to-peer topology that is highly capable of storing data on a global scale over thousands of participating servers. The technology has evolved far beyond its initial development for cryptocurrencies. Today, the technology has been implemented by various sectors and industries due to its ability to provide security to its immutable database over its decentralized network. This paper provides a detailed study into providing a feasible solution based on blockchain technology for the real-time authentication and secure deployment of hashed academic credentials issued by higher institutions in Brunei Darussalam under the governing body of the sultanate's Ministry of Education. The feasibility of the designed system is tested through the execution of smart contracts onto the Kovan Ethereum test network.

Keywords: Blockchain · Distributed ledger · Decentralized · Immutability · Education · Academic credentials · Students · Employers · Brunei darussalam

1 Introduction

Blockchain technology has often been associated with cryptocurrencies such as Bitcoin (BTC) since its initial development by Satoshi Nakamoto to be used as a digital peer-to-peer electronic cash system [1]. Underwood argued that blockchain technology has capabilities that transcend past cryptocurrencies and has revolutionized application developments and redefine the digital economy on a global scale [2]. The author further describes the technology as an immutable, fast, transparent, reliable, and trustworthy solution for various sectors and use cases in developing countries [2].

Brunei Darussalam has articulated a growing interest in becoming a smart nation as the sultanate intends to meet the standards of the "Wawasan Brunei 2035" vision which aims to raise the national standards of education by producing highly educated and skilled workforce through the development of a modern economy that will be able to provide exceptional employment opportunities for the nation's future leaders. Blockchain technology has the ability to enhance an individual's employability by providing instant authentication of academic credentials issued onto the blockchain by education institutions under the purview of the Ministry of Education. This will act as a trustworthy platform for employers both locally and internationally in the verification process for academic certificates without the need for intermediaries through smart contracts.

W. S. H. Suhaili et al. (Eds.): CIIS 2020, AISC 1321, pp. 75–84, 2021.
https://doi.org/10.1007/978-3-030-68133-3_8

2 Literature Review

2.1 Blockchain Technology

Crosby, Pattanayak, Verma, and Kalyanaraman describes blockchain technology as a distributed database of records or public ledgers containing transactions or executed digital events verified by a consensus of a participating majority of nodes on a peer-to-peer decentralized network [3]. The technology's greatest advantage is its ability to manage digital assets by validating transactions, securing entries, and preserving historic records added to the blockchain for future auditing [3].

Zheng, Xie, Dai, Chen, and Wang identified three of the most utilized types of blockchain: the public blockchain, consortium blockchain, and private blockchain. Each type provides a different permission type, level of immutability, level of efficiency, governance, and consensus process, as shown in Table 1 [4]. The consensus determination involves the nodes that validate and verify the blocks, the reading permission determines the access limitation, the immutability determines the ability to ensure that the data stored on the blockchain is not maliciously modifiable, the efficiency is based on the time taken to propagate transactions and blocks, centralization is the ability to control the blockchain, and consensus process is the limitation of allowing nodes to participate on the network [4]. It is virtually impossible to tamper with a public blockchain given the number of nodes it has on its network, the more nodes there are, the more computing power required to maliciously make changes to the majority of the network. The public blockchain is fully decentralized where there is no single point of control over the network thus providing any node from participating permission-less.

Table 1. Types of blockchain [4]

Property	Public Blockchain	Consortium Blockchain	Private Blockchain
Consensus Determination	All Miners	Assigned Nodes	One Organization
Read Permission	Public	Public or Restricted	Public or Restricted
Immutability	Virtually Impossible	Tamper-prone	Tamper-prone
Efficiency	Low	High	High
Centralization	No	Partial	Yes
Consensus Process	Permission-less	Permissioned	Permissioned

Xu, Tian, and Li identified five core concepts that are fundamental and essential features of blockchain technology: the block, timestamp, hash algorithm, asymmetric cryptography, and the Merkle tree structure [5]. Bhaskar, Devi, and Chuen state that blocks contain multiple valid transactions that have been hashed and encoded onto a Merkle tree structure containing a cryptographic hash of the previous block forming a chain of blocks [6]. As soon a block is generated on the blockchain, a timestamp for the block is created and it matches the authentication of every transaction to maintain authenticity. The timestamp shows that the block is organized in chronological order and marks the time for every transaction that acts as a notary [5].

Singhal, Dhameja, and Panda describe hash algorithms as the mathematical cryptographic functions that are fundamental in a blockchain data structure involving various cryptographic protocols, information security applications such as digital signatures and message authentication codes converting data to fixed-length hashed values [7].

Asymmetric cryptography involves the use of a public key and a private key as shown in Fig. 1 where the data is first converted into a hashed value which will then be encrypted using the signer's private key to generate a digital signature whereas the signer's public key would be required to decrypt the digitally signed data to be compared with the data provided for validation if the hash value matches accordingly.

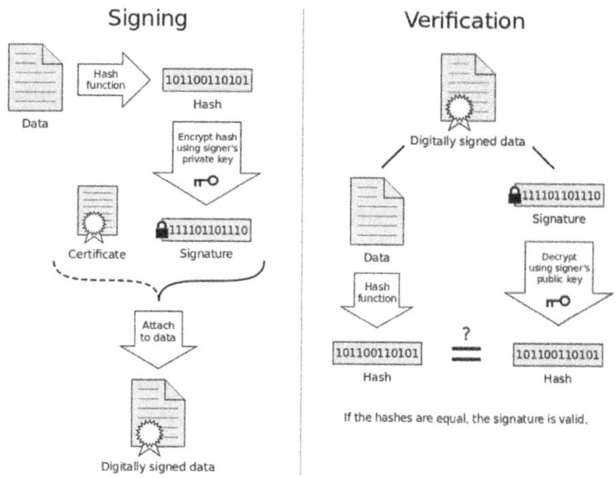

Fig. 1. Digital signature asymmetric cryptography [8]

Garner describes the Merkle tree as a structure that maintains and organizes multiple transactions to produce one hash for each block on the blockchain. If any changes are made to any one datum within the structure, it will modify the final result of the Merkle root generated from the hashed data as shown in Fig. 2 [9].

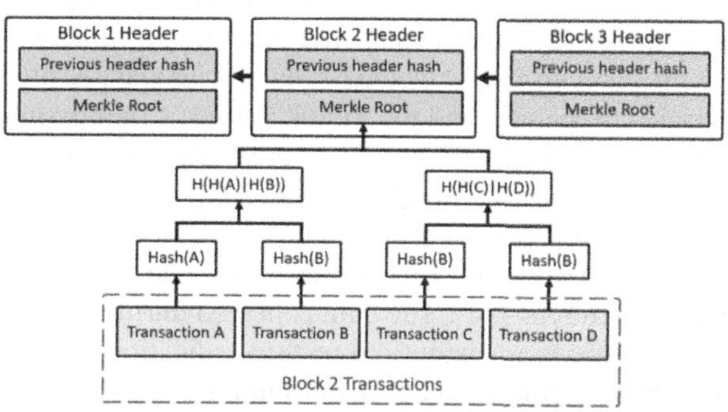

Fig. 2. Merkle tree structure [10]

2.2 Brunei's Current Education and Recruitment System

The government of Brunei Darussalam provides full funding for the education of its citizens to leverage the opportunity to advance knowledge [11]. The Compulsory Education Act states that Bruneian citizens that have reached the compulsory school age of seven years old are required by law to receive education [12]. According to the Brunei Darussalam Education Statistics and Indicators, the nation currently has 251 educational institutions with approximately 110,057 students and 10,686 teachers [13].

The Ministry of Education established the Brunei Darussalam Accreditation Council to ensure that academic credentials are up to the national standards and provisions required by the government by assessing and ascertaining the academic qualifications. Private institutions are required to obtain accreditation prior to delivering new courses suggesting that the system will be controlled by the ministry, independent government institutions are provided with pre-approval to issue academic credentials on the system, and private institutions to acquire permission from the council as shown on Table 2.

Table 2. Controlling structure of institution accreditation

Ministry of Education	
Independent Government Institutions	Brunei Darussalam Accreditation Council
Universiti Brunei Darussalam	Laksamana College of Brunei
Universiti Teknologi Brunei	International Graduate Studies College
Universiti Islam Sultan Sharif Ali	Kemuda Institute
Kolej Universiti Perguruan Ugama Seri Begawan	Cosmopolitan College of Commerce and Technology
Politeknik Brunei	Micronet International College
Institute of Brunei Technical Education	BICPA-FTMS Accountancy Academy

The Brunei Darussalam Accreditation Council disclosed certain cases and issues their unit confront in authenticating fraudulent and unverifiable academic credentials submitted for accreditation. The representative declared that it is also a requirement for the Public Service Commission applicants to acquire accreditation from the council as part of the government's strict recruitment procedure. A major concern would be the current authentication process carried out by the Brunei Darussalam Accreditation Council, recruiters from private companies, and the Public Service Commission where physical academic credentials are reviewed in person, whereas the council would be limited to checking their system for GCE 'O' and 'A' Level results. According to Contreras and Gollin, fake diploma mills gain illegal profit from issuing fake academic credentials that are equipped with security paper and seals that are almost identical to the original copies issued by legitimate institutions [14]. The data collected during the research states that the average time of response from institutions on the verification of academic credentials would take a week which is considered time-consuming.

Another matter shared by stakeholders and research participants is the numerous paper copies issued to higher institutions, scholarship units, employers for job applications which eventually contributes to storage space and paper consumption over time.

2.3 Blockchain Technology for Academic Credentials Authentication

According to Grech and Camilleri, developing smart cities such as Malta, Estonia, and most recently, Singapore, have already begun integrating their education system with blockchain technology [15]. The use of blockchain technology is utilized for the varied authentication methods of academic certificates and to act as a secure repository to store records of digital academic certificates legitimately issued.

In 2019, GovTech Singapore launched OpenCerts onto the blockchain in which the organization describes is an open-source schema for broadcasting educational certification through the use of a set of cryptographic tools to encrypt and protect the academic credentials, and an online website for the public to verify and authenticate certificates produced by OpenCerts [16]. Their system allows educational institutions to issue digital certificates through the platform for their students. The developers stated that blockchain technology on the Ethereum network "will greatly reduce the barrier to entry of publishing cryptographically protected educational credentials."

The Malaysian Ministry of Education launched the eScroll system in 2018 based on the university consortium on blockchain technology on a joint venture with Dagang NeXchange Berhad (DNeX), an award-winning company in providing eServices, and LuxTag, a solution provider for anti-counterfeit and anti-theft [17]. The ministry's goal is to provide a widespread implementation of the system, skills, training, and development of blockchain technology into their nationwide education sector where the university level has already been declared a priority. This would eliminate the time and energy-consuming processes of managing over thousands of manual verifications carried out over telephone calls and e-mails globally.

Based on an article written on "LuxTag's New Solution e-Scroll to Revolutionize Educational Certificates," eScroll is a solution for proof of academic certification ownerships and anti-counterfeiting [18]. The article further explains that due to the selling of fake diplomas on the black market, it has been alarming and harmful to industries such as the health sector which requires professional medical certification [18]. The developers claim that the system is sage, fast, and easy to operate [18]. QR codes will be placed on every certificate issued, which is linked to the blockchain, having features such as security, immutability, and an incorruptible database of academic records. The developers chose blockchain due to its ability to operate without a central point of authority like regular databases, the technology also has a non-destructive and immutable technique to track changes in data over a period [18].

Grech and Camilleri stated that The Ministry of Education and Employment (MEDE) in Malta believes that the implementation of blockchain technology would be able to provide learners and workers with maximum ownership and mobility of their own academic credentials through self-sovereignty [15]. The citizens of Maltese now own their academic credentials as an effective and skilled member of the modern workforce that is driven by mobility, international, and self-developing lifelong learners [15]. Brunei Darussalam can leverage the implementation of blockchain technology and its advanced cryptographic security advantages in its education system to ensure that students, institutions, employers, and scholarship units are able to effectively authenticate academic credentials onto the blockchain without fear of malicious attacks.

3 Proposed System

3.1 System Workflow Design

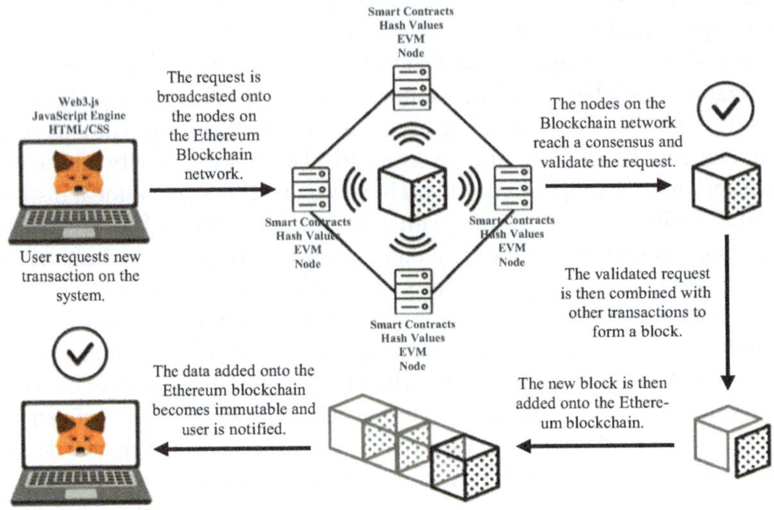

Fig. 3. System workflow design

Figure 3 demonstrates the system workflow design in the instance that a user adds a new transaction onto the Ethereum blockchain. The workflow begins when the user requests a new transaction which will require a minimum amount of Ethers that can be managed in the MetaMask wallet. Once the Ethers for the transaction has been set, the transaction will then be broadcasted onto the blockchain peer-to-peer network of nodes. The nodes will then have to validate the transaction using an algorithm. Once the transaction has been verified, it will be combined with other transactions to form a block of data as a ledger. The new block containing the transaction will then be added to the previous block which will form a chain and become immutable. The transaction will then be completed, and the issuer will be notified. This applies to both the issuance of permission to institutions onto the system and also the issuance of academic credentials onto the blockchain. The speed of the transaction is also influenced by the Ether, the higher the Ether, the faster the transaction will be validated and completed.

The role of the Ethereum blockchain is to store the transactions of hash value generated and converted by uploading the PDF academic credentials running on the Ethereum Virtual Machine (EVM). The PDF academic credentials are converted using Keccak-256, a cryptographic function generally used for Ethereum. The logic or application code is more commonly known as the smart contract, in this case, Solidity. It will be compiled using Remix IDE to be deployed onto the Ethereum test network. Web3 is a JavaScript library that allows the browser to interact with the Ethereum blockchain. JavaScript frameworks involved are ReactJS and NodeJS.

3.2 User Workflow Design

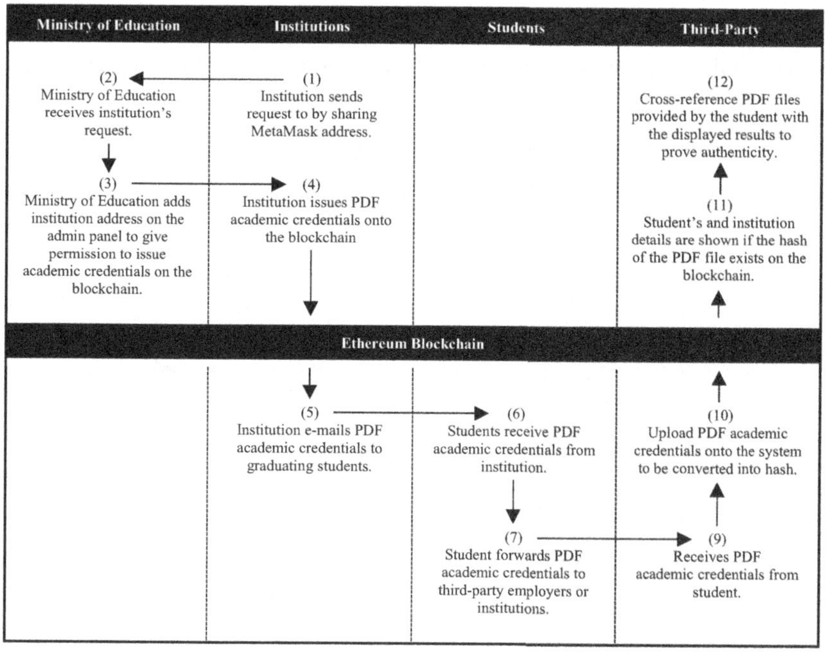

Fig. 4. User workflow design

The user workflow design shown in Fig. 4 is constructed based on the simplification of existing authentication processes carried out in Brunei Darussalam that will be achieved in the proposed system. The figure demonstrates that the institution will first have to send a request to the Ministry of Education by sharing the necessary details and the MetaMask address. If the institution meets the criteria set by the Brunei Darussalam Accreditation Council (BDAC), the institution's MetaMask address will be added to the admin panel along with the details of the institution. This will give the institution's address permission to be directed onto the issuer's panel where they can issue academic credentials onto the blockchain. The institution will upload the PDF file onto the system which will be converted into a fixed-length hash value which will be added onto the blockchain along with the student details. Upon a successful transaction, the institution may send the PDF files to the graduates. The students will then receive the PDF files via e-mail and have the option of forwarding it to third-party employers, higher institutions, scholarship units, etc. The third-party verifiers will then upload the PDF files onto the system which will again be converted into a hash value in which the system will cross-reference with the data stored in the blockchain. Should the hash value exist, then the results matching the value will display with the issuing institution details and the student's details. If the hash value does not exist, this would mean that the credentials do not exist in the system or the PDF file has been modified.

4 Implementation

4.1 Smart Contracts

The solidity smart contracts shown in Table 3 are deployed onto the Kovan test network to test the feasibility in the authentication of academic credentials on the blockchain.

Table 3. Smart contracts

No	Smart Contract	Description
1	`modifior onlyMinistryOfEducation { require(msg.sender == ministryOfEducation, "Only the Ministry Of Education can run this function."); _; }`	Sets the Ministry of Education's as the only address that can make modifications to permissions and additions for issuers.
2	`function create(string memory name, address adr) public onlyMinistryOfEducation { require(adr != address(0), "Address is equal to zero."); uint id = institutions.length; institutions.push( Institution(id, adr, name, true)); isApproved[adr] = true; }`	Function to add the institution's name and address to the Ethereum blockchain to allow access to the issuer panel.
3	`function approve(address adr) public onlyMinistryOfEducation { require(adr != address(0), "Address is equal to zero."); require(!isApproved[adr], "Address is already approved."); uint index = getInstitutionIndex(adr); institutions[index].isApproved = true; isApproved[adr] = true; }`	Function to approve institution address to allow issuance of academic credentials onto the Ethereum blockchain.
4	`function disapprove(address adr) public onlyMinistryOfEducation { require(adr != address(0), "Address is equal to zero."); require(isApproved[adr], "Address is not approved."); uint index = getInstitutionIndex(adr); institutions[index].isApproved = false; isApproved[adr] = false; }`	Function to disapprove institution address. This will not redirect the institution to the issuer panel but will only display the verify function.
5	`function update(address ministry) public onlyMinistryOfEducation { require(ministry != address(0), "Address is equal to zero."); ministryOfEducation = ministry; }`	Updates the address of the Ministry of Education, transfer of ownership of the smart acontract controlling the permissions.
6	`function addRecord(string memory studentID, string memory studentName, string memory programme, string memory result, string memory modeOfStudy, string memory hash, string memory date) public { require(isApproved[msg.sender], "Address is not approved."); require(!isHashExist(hash), "Hash already exists."); records.push(Record(studentID, studentName, msg.sender, programme, result, modeOfStudy, hash, date)); }`	Adds record of the Student ID, Student Name, Institution Name, Programme, Result, Mode of Study, Date Issued, and generated hash value of PDF file.
7	`function getRecord(string memory hash) public view returns (Record memory) { for (uint i=0; i<records.length; i++) { if (keccak256(abi.encodePacked( (records[i].hash))) == keccak256(abi.en-codePacked(hash))) { return records[i]; } } }`	Matches generated hash value of the PDF file uploaded and stored in the Ethereum blockchain with the generated hash value of the PDF file to be verified.
8	`function isHashExist(string memory hash) public view returns (bool) { } for (uint i=0; i<records.length; i++) { if (keccak256(abi.encodePacked( (records[i].hash)))==keccak256(abi.encodePacked(hash))) { return true; } } return false; }`	Checks if generated hash value using Keccak-256 of the academic credentials exist in the Ethereum blockchain.
9	`function getInstitutionIndex(address adr) public view returns (uint) { for (uint i=0; i<institutions.length; i++) { if (institutions[i].adr == adr) { return institutions[i].index; } } }`	Displays issuer panel for authorized institution address by the Ministry of Education.

4.2 Black Box Functionality Testing

Black box testing is conducted on the Kovan Ethereum Test Network in a MacOS development environment requiring software such as Chrome, MetaMask, HomeBrew, Xcode, Go Ethereum, Ganache, NodeJS, NPM, Truffle, Atom, and Remix IDE.

Table 4. Black box functionality test results

No	Function tested	Expected results	Results
1	Set MetaMask account to main Minister of Education account	Displays permission panel for Ministry of Education	Success
2	Set MetaMask account to issuer institution address	Display issuer panel for institutions with permission	Success
3	Set the MetaMask account to a random address	Displays the verification page only	Success
4	Submit institution address and institution name while logged in as Minister of Education MetaMask account	Institution permission is added to the Ethereum blockchain and a success message will be shown from MetaMask	Success
5	Revoke institution by clicking "Disapprove" button while logged in as Minister of Education MetaMask account	Permission for the institution is revoked and a success message will be shown in MetaMask. Institution will no longer be able to access Issuer Panel	Success
6	Re-approve institution by clicking "Approve" button while logged in as Minister of Education MetaMask account	Permission for institution is reinstated and success message will be shown in MetaMask Institution will be able to access Issuer Panel	Success
7	Approved MetaMask institution address submits PDF academic credentials and all relevant details	Hash value of converted PDF file and relevant details is issued onto the blockchain and MetaMask will display success message	Success
8	Approved MetaMask institution address submits duplicate PDF academic credentials and details	An error message will display	Success
9	Uploads existing PDF academic credentials on the Ethereum blockchain to be verified	Relevant details will be successfully displayed	Success
10	Uploads modified PDF files or have not been issued by the institution onto the system	Shows message that the credential is not verifiable	Success

Figure 5 below displays the successful verification of an academic credential uploaded where the hash value of the PDF file has not been modified and exists in the blockchain where the details of the issuing institution and students associated with the credentials will be displayed. Should the hash value not exist where the PDF file has been modified or forged, an error message display stating that the file uploaded is not verifiable.

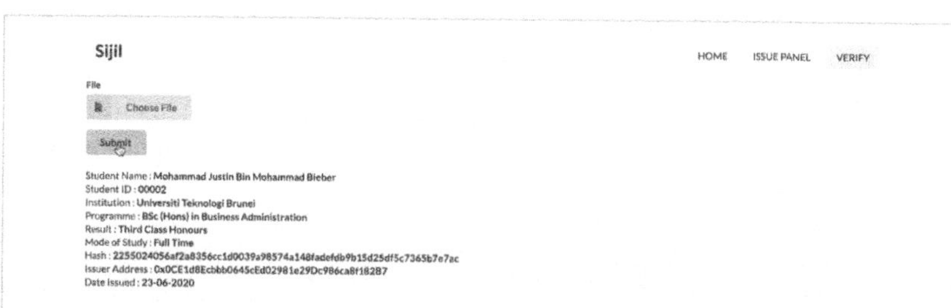

Fig. 5. Successful verification

5 Conclusion

This study concludes that blockchain technology has capabilities in providing a trustworthy, secure, and more convenient alternative with regards to the real-time authentication of academic credentials issued onto the blockchain. The implementation of the technology in Brunei Darussalam's education sector is proven to be feasible but is still limited by the current process where lower institutions that obtain certifications internationally issued by the University of Cambridge poses a challenge. Future recommendations would be to allow accredited international institutions to issue academic credentials onto the same system upon recognition by the Brunei Darussalam Accreditation Council. It is essential that institutions within the nation elect a trustworthy user to operate the system as data issued onto the blockchain is irrevocable. Regardless, the system can confidently provide both local and international users with convenience.

References

1. Nakamoto, S.: Bitcoin: A Peer-to-Peer Electronic Cash System. Bitcoin, Virginia (2008). https://bitcoin.org/bitcoin
2. Underwood, S.: Blockchain beyond bitcoin. Commun. ACM **59**(11), 15–17 (2016)
3. Crosby, M., Pattanayak, P., Verma, S., Kalyanaraman, V.: Blockchain technology: beyond bitcoin. Appl. Innov. **2**(6–10), 71. Berkeley, California (2016)
4. Zheng, Z., Xie, S., Dai, H., Chen, X., Wang, H.: An overview of blockchain technology: architecture, consensus, and future trends. In: 2017 IEEE 6th International Congress on Big Data, pp. 557–564 (2017)
5. Xu, M., Tian, Y., Li, J.: Blockchain. Skyhorse Publishing, La Vergne (2018)
6. Bhaskar, N. D., Chuen, D. L. K.: Bitcoin Mining Technology. Handbook of Digital Currency, pp. 45–65. Academic Press, Massachusetts (2015).
7. Singhal, B., Dhameja, G., Panda, P.S.: Beginning Blockchain: A Beginner's Guide to Building Blockchain Solutions. Apress, New York (2018)
8. Digital Signature. https://w.wiki/mt4. Accessed 20 Oct 2020
9. Garner, B.: Merkle Tree Hashing: How Blockchain Verification Works (2018). https://coincentral.com/merkle-tree-hashing-blockchain/
10. Chen, Y.C., Chou, Y.P., Chou, Y.C.: An image authentication scheme using merkle tree mechanisms. Future Internet **11**(7), 149 (2019)
11. Johnson, B.: Free Education in Brunei Really Aids Nation's Growth (2017). https://borgenproject.org/free-education-in-brunei/
12. Compulsory Education. Attorney General's Chamber, Brunei Darussalam (2011). https://www.agc.gov.bn/AGC%20Images/LAWS/ACT_PDF/cap211.pdf
13. Education Statistics and Indicators. Ministry of Education, Brunei Darussalam (2018). https://www.moe.gov.bn/DocumentDownloads/Education%20Statistics%20and%20Indicators%20Handbook/Brunei%20Darussalam%20Education%20Statistics%202018.pdf
14. Contreras, A., Gollin, G.: The Real and The Fake Degree and Diploma Mills. Change: The Magazine of Higher Learning, 41:2, 36–43. Council for Higher Education Accreditation, Washington (2009). https://doi.org/https://doi.org/10.3200/CHNG.41.2.36-43
15. Grech, A., Camilleri, A.F.: Blockchain in education. Publications Office of the European Union, Luxembourg (2017). http://dx.doi.org/https://doi.org/10.2760/60649
16. OpenCerts. https://opencerts.io/. Accessed 20 Oct 2020
17. LuxTag. https://link.medium.com/USRZDwiBtbb. Accessed 20 Oct 2020
18. Lacassagne, C.: LuxTag's New Solution e-Scroll to Revolutionize Educational Certificates (2019). https://link.medium.com/vDn010pBtbb

Privacy Policy, Training and Adaption of Employee Monitoring Technology to Curtail Workplace Harassment in Organizations: An Application of TAM

Zahidah Ismail and Shahid Anjum(✉)

School of Business, Universiti Teknologi Brunei, Gadong BE1410, Brunei
zahidahismailothman@gmail.com, shah.jum@utb.edu.bn

Abstract. Digital technology is becoming more sophisticated and so are organizational monitoring technologies implemented to enhance business efficiencies. Some employees may accept such changes effortlessly, but some will resist to adopt to the digitization and monitoring technologies' implementation because of several concerns including the concerns of privacy violation. This gives rise to a research question about the adaption of the monitoring technologies that is the main theme of this research. A survey of various categories of employees has been conducted in Brunei's economy using a questionnaire, designed based on Technology Adaption Model has been disseminated widely. The response rate was fair, however, 17% of all responses have been rejected based on various deficiencies and only 179 responses have been used for analysis. The study has tried to find the answer about the behavioral intention to use the employee monitoring technology by employees by considering various organizational, contextual and individual factors. Data analysis has used Cronbach's alpha, correlation, and linear regression. Questionnaire consistency has been found to hold. In order to establish the statistically significant causation in the proposed model hypothesis, few iterations are required. Privacy policy and training of employees enhance understanding of technology. Perceived usefulness, which is affected by ease of use but not directly from the understanding variable. Subjective norms variable has an impact on intention to use where the main determinant of usage behavior, which has come up statistically significant, is workplace harassment reduction feature of the monitoring technology.

Keywords: Technology acceptance model · Employee monitoring technology · Organizational efficiency · Workplace harassment · Brunei darussalam

1 Introduction

Employee Monitoring Technology (EMT) is a system installed in the employees' computer with the purpose to monitor the employees' activities [19]. By using EMT, the employer will be able to monitor employees simultaneously and therefore may enhance efficiency and reduce the unethical behavior which may include, to the extreme, fraud,

W. S. H. Suhaili et al. (Eds.): CIIS 2020, AISC 1321, pp. 85–94, 2021.
https://doi.org/10.1007/978-3-030-68133-3_9

and misuse of resources [27]. Frauds are unethical behaviors by dishonest individuals to obtain benefits in unacceptable way [24]. It is stated that organization that reported no indication of fraud is most likely because the organization only had fewer controls compared to organization that had more controls over their organization [22]. In other words, although there is no fraud reported in the organization, but it does not mean that the organization is free from fraud. Hence, it is important for the organization to ensure that the employees to do their task by following the correct procedure and rules [20].

Apart from implementing EMT to combat fraud, there are other ways which includes password protection and firewall protection [6]. However, several other issues have arisen by the employees regarding the adaption of the EMT such as the feeling of being untrusted by the management employees, being unfair and abusive [3]. Not being able to trust is also one of the negative feelings which can damage the trust relationship between the employer and the employees [8]. Employees who have unlimited and unrestricted access to internet has caused more risk to breaching [11]. By implementing the EMT in the workplace, private boundary turbulence will occur because both employer and employee seizing the ownership of the information about employees [8]. In order to avoid ethical issues, organizations should be considering the employees' feelings and should be able to balance by monitoring the employees' activities in acceptable way [16]. Employees should also be given a training regarding the EMT [22]. There are several proposed recommendations before implementing the EMT which includes identifying the purpose of monitoring, obtaining the consent from the employees, providing the employees with the policy regarding the use technology, informing the employees how and when will the employees be monitored, as well as informing the employees that password and firewall are not enough to prevent fraud [4]. Although many employees do not agree with the implementation of EMT due to several concerns including that EMT that may invade the employees' privacy, however they may not understand that it may have the tendency to protect their personal information [26].

Therefore, this study has been conducted to understand the adaption behavior of the employees regarding employee monitoring technology using Technology Acceptance Model where the reduction of workplace harassment has been used as one of the determinants for intention to use employee monitoring technology. This paper is divided into six sections: introduction, literature review, methodology, result and findings, discussion and limitation of study, and conclusion.

2 Literature Review

Technology like EMT enable the employer to observe on employees' activities in the workplace which includes capturing employees' productivity and reviewing employees' communications via email and internet [19]. The main purpose of EMT is to monitor employees' productivity, logistic tracking and for employees' safety [26]. There are few ways to monitor the activities of the employees which includes email audit, counting the keystrokes, record time during online, and view computer screen in real time [15]. EMT is a technology that can be used to collect data of employees' activities which includes keystroke, attendance, time taken to finish certain tasks, and time taken for break time. The data that is collected by EMT can be utilized by organization for monitoring where the data can be used for improvement by taking actions for employees, who has taken more time to complete a task compared to other employees, in order to retrain them. However, before implementing EMT in the workplace, the organization should consider employees that may resists to the implementation of the EMT for several reasons.

Employer's perception on EMT may not the same as employees' perception [19]. It also has been stated that employer and employees may have different perspective and opinions with the implementation of EMT [16]. In other words, the factors of why employees resist to the implementation of EMT relates to the perception of the employees toward EMT. Employees may feel that the employer has no trust towards employees and employees may feel like their privacy has been invaded. On the other hand, trust consists of trustor and trustee where the trust can be gained by transparent communication [28].

The organization should consider what type of monitoring is acceptable and what type of activities are prohibited in the workplace [27]. Although the organization has rights to implement EMT, but the organization should not violated the employee right of privacy as employees should be informed that they will be monitored, how will they be monitored, how the data will be stored, and confidentiality of the data [17]. Employees may have different perceptions on being monitored as they may against with the idea of them being monitored at the workplace [8]. Employees will normally give negative feedback with the idea of employees being monitored as employees will be feeling stress, injustice and dissatisfaction [25].

There have been many fraudulent cases being reported in organization in Brunei Darussalam. A local newspaper has reported a fraudulent case which has siphoned Brunei Dollars 84,928.90 in between 7th August and 22nd December, 2009 where the suspect was an assistant branch manager of a bank, quite high position to have the privilege to access the confidential data [12]. Fraud is irresponsible acts by dishonest individuals to obtain benefits in unethical ways which is very unacceptable [5]. If the organization did not take any action and ignore the fraudulent cases, the impact can be big towards the organization and even to the country.

The implementation of EMT may cause the employee to have a feeling of privacy invasion and unfairness. Nobody wants their privacy being invaded by anybody especially when it used for defamation. Thus, this is considered as privacy infringement [26]. Despite people have been posting contents to social media voluntarily, but people still do not want any people to invade their privacy. On the other hand, if EMT is implemented to the organization without the employees' consent, the employees may feel that the organization have invade their privacy and they may feel that their privacy have been infringed. There are also an argument that stated EMT has not affected the employees' behavior in the workplace as some employees also perceive that behavioral surveillance is important to create a sense of security for employees and the public because employees believe that it is inevitable in this modern era to be recorded and tracked [20]. Organization should implement EMT with policies and principle for fair and lawful monitoring [16].

Privacy policy is a document where it describes many important things which also includes telling the users what type of data will be collected and to which third party will the data be shared [1]. Therefore, software or application must provide their users a privacy policy. It can be a serious matter if the user does not bother to read the privacy policy [21]. If the software company can properly store and analyzed the collected data, the data can be used to understand the customers' behavior [9]. If the employees fully understand the privacy policy, the employees may have focus on a similar mission [10]. In addition to this, employees will perceive that EMT can be useful and easy to use if the organization provides the employees a proper explanation on the privacy policy and provides training for the employees.

This research utilises a model for evaluating the acceptance of a system entitled Technology Acceptance Model (TAM) as the main basis for developing the research model. TAM has been widely used to study the consumers' behavioral intention to use or to study consumers' interest and perception of a new technology [14]. TAM has been used widely due to its simplicity to measure the acceptance of consumers towards the new technology [2]. TAM can be defined as a model that shows how a technology can be accepted by its intended users, which mostly affected by their attitudes and perspectives and there are four major components in TAM, namely the Perceived Ease of Use (PEOU), Perceived Usefulness (PU), Attitude towards Usage (ATU), and Behavioral Intention to Use (BIOU) [18] where the model is used for this paper. Acceptance in this model means a positive feedback from the consumers to use the new technology [23].

3 Methodology

The proposed variant model of TAM as eight components which are privacy policy (PP), training (T), understanding of EMT (UNDR), perceive of usefulness (POU), ease of use (EOU), subjective norm (SOC), intention to use (ITU), and usage behavior (UB) (see Fig. 1). The existence of PP in the organization for handling confidential information. Training for handling confidential information is provided by the organization is represented by T. SOC represents the group of influencers who influence the employees to use EMT. ITU is an average of the three variables which has been labelled as ITU1, ITU2, and ITU3 that can be described as the increase productivity of organization, increase confidence of customers of organization, and decrease of unnecessary harassment at workplace, respectively. UB represents usage behavior of the employees toward the EMT. Likewise, other variables have been divided into differentiable categories – PP into PP1 and PP2, T into T1 and T2, and SOC into SOC1, SOC2, and SOC3. SOC1, SOC2, and SOC3 can be defined as family influence, department influence, and government influence, respectively. There are four items in PP1 and two items in PP2 to formulate an aggregate variable named PP. Both T1 and T2 have two items formulate an aggregate variable named T.

Various hypotheses (represented by prefix H) of hypothesized proposed model have been shown in Fig. 1. The existence of privacy policy in handling confidential information in organization can lead into UNDR, POU, and EOU which is represented by H1, H2, and H3, respectively with the assumption of positive correlation for all three cases. Likewise, training that is provided by the organization for the employees in handling confidential information of the organization will have positive impact with UNDR, POU, and EOU which is represented by H4, H5, and H6, respectively with the assumption of positive correlation for all three cases. Moreover, UNDR, POU, EOU, and SOC will lead the employees to have intention to use the technology which is represented by H7, H8,

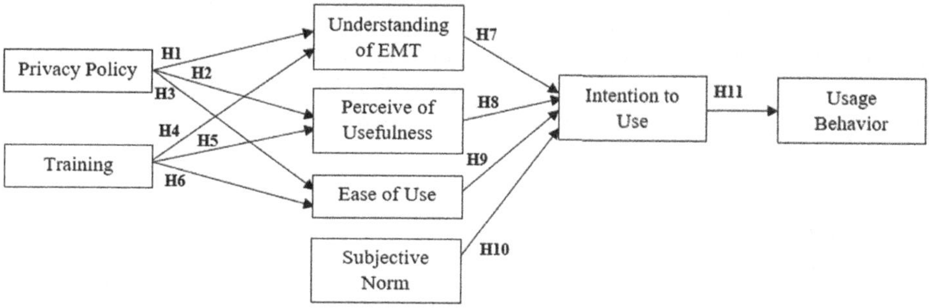

Fig. 1. Proposed model

H9, and H10, respectively with the assumption of positive correlation for all four cases. The relationship of ITU and UB is represented by H11 with the assumption of positive correlation.

The proposed model has needed few iterations before an empirical model has been found where all variables are significant. UNDR has supported POU indirectly via EOU, and EOU has supported ITU indirectly via POU as shown in the Empirical Model (see Fig. 2).

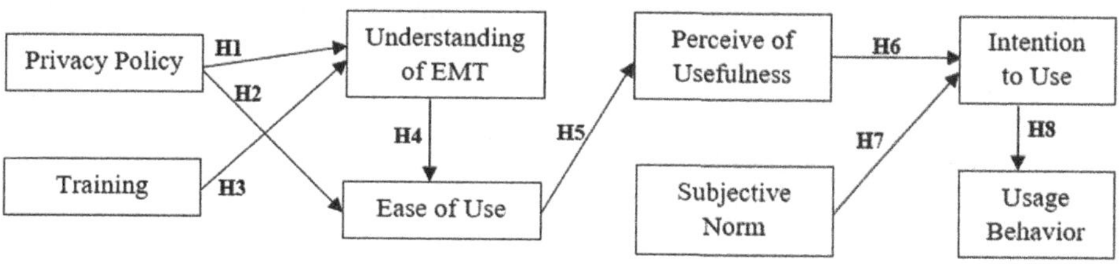

Fig. 2. Empirical model

4 Result and Findings

4.1 Sample Characteristics

This study employs a quantitative research, where a questionnaire has been circulated to the employees, using convention and digital means in Brunei Darussalam. The questionnaire was divided into 6 sections; demographic, organization culture of policies, ease of use, perceive of usefulness, subjective norm, and understanding of EMT. Demographic section includes gender, age, marital status, highest level of education, type of employment, professional area of employment, and income. Each question (items) from all sections except for the demographic section has been measured through 6-point Likert scales of strongly agree, moderately agree, simply agree, simply disagree, moderately disagree, and strongly disagree.

Three statistical methods have been used named as Cronbach's alpha, correlation, and linear regression by using SPSS software in order to perform a total of 215 responses have been collected but only 179 responses are complete and can be used for data analysis. The results of Cronbach's alpha and linear regression have been presented in Table 1 through Table 2.

4.2 Results

Cronbach's Alpha has been used to examine the reliability of the data in this study as is shown in Table 1 where all of those are significant. The study revealed that all the correlations are positive except for; PP2 with all variables. T, T1, and T2 with SOC, SOC1, SOC2, SOC3, and UB, respectively across all variables. The correlations that are less than 0.5 except for, PP with PP1, T, and T1. PP1 with T and T1. T with T1 and T2. T1 with T2. POU with EOU, ITU, ITU1, ITU2, and ITU3. EOU with ITU, ITU1, ITU2, and ITU3. SOC with SOC1, SOC2, and SOC3. SOC1 with SOC2 and SOC3. SOC2 with SOC3. ITU with ITU1, ITU2, and ITU3. ITU1 with ITU2 and ITU3. ITU2 and ITU3.

Table 1. Cronbach's alpha results for dataset.

Variables	Cronbach's alpha	Variables	Cronbach's alpha	Variables	Cronbach's alpha
PP1	0.910	UNDR	0.789	SOC2	0.987
PP2	0.906	POU	0.954	SOC3	0.982
T1	0.995	EOU	0.993	ITU	0.911
T2	0.995	SOC1	0.921		

Various linear regression has been tested several times with different variables separately and simultaneously with different combination to find statistically significant relation which has landed us at the empirical model whose results have been shown in Fig. 2 and where all variables which have supported tested hypotheses have been shown the arrows of all are statistically significant as are presented in Table 2. Based on the result, all hypotheses in Fig. 1 are significant except for H5, H6, H7, and H9. There are four hypotheses were not supported, therefore multiple times of iterations of linear regressions have been made to test the significancy. After several times of iterations where all them are statistically significant, it has shown that UNDR supported POU indirectly via EOU and it also has shown that EOU supported ITU indirectly via POU. SOC1, SOC2, and SOC3 were tested whether these three variables have a positive impact towards ITU or not. Based on the result, it has shown that SOC3 significantly have a positive impact towards ITU. ITU1, ITU2, and ITU3 were also tested to see whether these three variables have a positive impact towards UB or not. Based on the result, it has shown that ITU3 is significantly has a positive impact towards UB.

Table 2. Linear regression results for dataset

Independent Variable	Adjusted R^2	Beta coefficient value	T-Value (T significant)
Regression 1: Dependent Variable: UNDR			
PP	0.162	0.345	4.357(*)
T		0.114	1.432
Regression 2: Dependent Variable: UNDR			
PP1	0.209	0.469	5.521(*)
T1		−0.284	−2.098(**)
T2		0.301	2.462(**)
Regression 3: Dependent Variable: EOU			
PP	0.200	0.452	6.738(*)
Regression 4: Dependent Variable: EOU			
PP1	0.247	0.384	5.164(*)
PP2		−0.193	−2.599(**)
Regression 5: Dependent Variable: EOU			
UNDR	0.200	0.452	6.740(*)
Regression 6: Dependent Variable: POU			
EOU	0.614	0.785	16.842(*)
Regression 7: Dependent Variable: ITU			
POU	0.860	0.905	30.522(*)
SOC		0.067	2.247
Regression 8: Dependent Variable: ITU			
POU	0.862	0.905	31.078(*)
SOC3		0.077	2.650(*)
Regression 9: Dependent Variable: UB			
ITU	0.121	0.355	5.052(*)
Regression 10: Dependent Variable: UB			
ITU3	0.130	0.367	5.243(*)

Notes Table 2: * means that the value is 99% significant, ** means that the value is 95% significant, *** means that the value is 90% significant.

5 Discussion and Limitation of Study

The purpose of this study is to assess the perspective of employees on the implementation of an EMT. Based on the proposed model, the relationship between PP and UNDR, POU, and EOU are have been significant. It has also been proven by previous study that information in PP is very significant to influence users to use an application [7]. The proposed model also shows that UNDR and EOU have not supported ITU. In contrast to this, both POU and SOC support ITU, where SOC3 variable is significant. ITU is also tested in this model, where it was found that ITU supported UB. However, when we have tested each of the ITU variables separately only ITU3 is significant. Meanwhile, the empirical model where the model shows that UNDR has support POU indirectly via EOU and EOU has also support ITU indirectly via POU.

This study has several limitations. Firstly, the number of respondents in this study is relatively small because of time constraint. There are also lot of incomplete or inconsistent responses which have made the sample size even further smaller because of 17% rejections out of the total number of 215 responses and, therefore, only 179 responses have been used for analysis. Another limitation in this study is that broader factors from the effects of culture on the perspective amongst the employees may have been incorporated. Last but not the least, more sophisticated analytical methodologies may have been used in order to capture non-linear relationships as well as mediating and moderating effects.

6 Conclusion

The differences of proposed model and empirical model shows that various hypothesis need to be adjusted. This is normal for this study which does not have lot of presence in the existing literature. Our results show that the government verdicts, being considered as highly legitimate, influence the behavior of employees to adapt EMT in a significant way based on the evidence found about SOC3 which has influence on the employees in adapting EMT. Besides, the relationship between T with POU and EOU have been insignificant where the possible reason of this might be because of lack of training provided to the employees [13]. Based on other results we can deduce that the employees have the intention to use the technology because the employees perceived that the technology can be used to decrease unnecessary harassment at workplace. Sexual harassment, one of the forms of workplace harassment, is one of the crimes that happened in the workplace [4] and our study has shown that employees are willing to adapt EMT for the reason that it will help in decreasing the workplace harassment at the workplace.

References

1. Aïmeur, E., Lawani, O., Dalkir, K.: When changing the look of privacy policies affects user trust: an experimental study. Comput. Hum. Behav. **58**, 368–379 (2016). https://doi.org/10.1016/j.chb.2015.11.014
2. Al-Emran, M., Mezhuyev, V., Kamaludin, A.: Technology acceptance model in m-learning context: a systematic review. Comput. Educ. **125**, 389–412 (2018). https://doi.org/10.1016/j.compedu.2018.06.008
3. Alder, G.S., Noel, T.W., Ambrose, M.L.: Clarifying the effects of internet monitoring on job attitudes: the mediating role of employee trust. Inform. Manag. **43**, 894–903 (2006). https://doi.org/10.1016/j.im.2006.08.008
4. Ariss, S.S.: Computer monitoring: benefits and pitfalls facing management. Inform. Manag. **39**(7), 553–558 (2002). https://doi.org/10.1016/S0378-7206(01)00121-5

5. Baldree, J.: Fraud Risk Management: A Guide To Good Practice. Chart. Inst. Manag. Accountants
6. Bierstaker, J.L., Brody, R.G., Pacini, C.: Accountants' perceptions regarding fraud detection and prevention methods. Manag. Auditing J. **21**(5), 520–535 (2006). https://doi.org/10.1108/02686900610667283
7. Capistrano, E.P.S., Chen, J.V.: Information privacy policies: the effects of policy characteristics and online experience. Comput. Stan. Interfaces **42**, 24–31 (2015). https://doi.org/10.1016/j.csi.2015.04.001
8. Chang, S.E., Liu, A.Y., Lin, S.: Exploring privacy and trust for employee monitoring. Ind. Manag. Data Syst. **115**(1), 88–106 (2015). https://doi.org/10.1108/IMDS-07-2014-0197
9. Chang, Y., Wong, S.F., Libaque-Saenz, C.F., Lee, H.: The role of privacy policy on consumers' perceived privacy. Gov. Inform. Q. **35**(3), 445–459 (2018). https://doi.org/10.1016/j.giq.2018.04.002
10. Constantin, E.C., Baias, C.C.: Employee voice - key factor in internal communication. Procedia – Soc. Behav. Sci. **191**, 975–978 (2015). https://doi.org/10.1016/j.sbspro.2015.04.319
11. Eivazi, K.: Computer use monitoring and privacy at work. Comput. Law Secur. Rev. **27**(5), 516–523 (2011). https://doi.org/10.1016/j.clsr.2011.07.003
12. Faisal, F.: Former bank employee jailed for fraud. Borneo Bulletin (2017)
13. Gan, M.F., Chua, H.N., Wong, S.F.: Privacy enhancing technologies implementation: an investigation of its impact on work process and employee perception. Telematics Inform. **38**, 13–29 (2019). https://doi.org/10.1016/j.tele.2019.01.002
14. Guritno, S., Siringoringo, H.: Perceived usefulness, ease of use, and attitude towards online shopping usefulness towards online airlines ticket purchase. Procedia – Soc. Behav. Sci. **81**, 212–216 (2013). https://doi.org/10.1016/j.sbspro.2013.06.415
15. Ella, V.J.: Employee monitoring and workplace privacy law. In: Lewis, P.C.J. American Bar Association Section of Labor and Employment Law National Symposium on Technology in Labor & Employment Law, vol. 14 (2016)
16. Mitrou, L., Karyda, M.: Employees privacy vs. employers security : can they be balanced? Telematics Inform. **23**, 164–178 (2006). https://doi.org/10.1016/j.tele.2005.07.003
17. Moussa, M.: Monitoring employee behavior through the use of technology and issues of employee privacy in america. SAGE Open. **5**(2), 1–13 (2015). https://doi.org/10.1177/2158244015580168
18. Mugo, D., Njagi, K., Chemwei, B., Motanya, J.O.: The Technology Acceptance Model (TAM) and its application to the utilization of mobile learning technologies. Br. J. Math. Comput. Sci. **20**(4), 1–8 (2017). https://doi.org/10.9734/BJMCS/2017/29015
19. Samaranayake, V., Gamage, C.: Employee perception towards electronic monitoring at work place and its impact on job satisfaction of software professionals in Sri Lanka. Telematics Inform. **29**(2), 233–244 (2012). https://doi.org/10.1016/j.tele.2011.08.003
20. Sarra, R.: How Employees' Monitoring Perceptions Affect Organizational Trust: The Moderating Role Of Organizational Justice. Doctorate Dissertation (2016)
21. Steinfeld, N.: "i agree to the terms and conditions": (How) do users read privacy policies online? An eye-tracking experiment. Comput. Hum. Behav. **55**(Part B), 992–1000 (2016). https://doi.org/10.1016/j.chb.2015.09.038
22. Suh, J.B., Shim, H.S., Button, M.: Exploring the impact of organizational investment on occupational fraud: mediating effects of ethical culture and monitoring control. Int. J. Law Crime Justice **53**, 46–55 (2018). https://doi.org/10.1016/j.ijlcj.2018.02.003
23. Taherdoost, H.: A review of technology acceptance and adoption models and theories. Procedia Manuf. **22**, 960–967 (2018). https://doi.org/10.1016/j.promfg.2018.03.137
24. The Chartered Institute of Public Finance and Accountancy (CIPFA): Fraud Definition and Examples. Chart. Inst. Public Finance Account (2010)

25. Thiel, C.E., Macdougall, A.E., Bagdasarov, Z.: Big (Benevolent) brother: overcoming the drawbacks of employee monitoring through ethical administration. Organ. Dyn. **48**(2), 19–28 (2018). https://doi.org/10.1016/j.orgdyn.2018.08.001
26. Tomczak, D.L., Lanzo, L.A., Aguinis, H.: Evidence-based recommendations for employee performance monitoring. Bus. Horiz. **61**(2), 251–259 (2018). https://doi.org/10.1016/j.bushor.2017.11.006
27. Yerby, J.: Legal and ethical issues of employee monitoring. Online J. Appl. Knowl. Manag. **1**(2), 44–55 (2013)
28. Yue, C.A., Men, L.R., Ferguson, M.A.: Bridging transformational leadership, transparent communication, and employee openness to change: the mediating role of trust. Pub. Relat. Rev. **45**(3), 101779 (2019). https://doi.org/10.1016/j.pubrev.2019.04.012

Computational Intelligence and Its Applications

A Comparative Study of Supervised Machine Learning Techniques for Deceptive Review Identification Using Linguistic Inquiry and Word Count

Dinooshi Poornima Jayathunga(✉), R. M. Iranthi Shashikala Ranasinghe, and Ramashini Murugiah

Department of Computer Science and Informatics, Uva Wellassa University, Passara Road, Badulla, Sri Lanka
{poornima,iranthi,ramashini}@uwu.ac.lk

Abstract. In e-commerce websites enabling a facility to leave a review/feedback about the product is one of best practice given by developers; which has a significant influence on consumer's buying behavior. Meanwhile, sellers and manufacturers are investigating online reviews for decision making. Therefore, this facility was misused by generating fake reviews. Filtering out of untruthful information becomes essential in this modern era. The goal of this study is to find a robust supervised machine learning approach to identify deceptive reviews through a comparative study for the content-based feature called Linguistic Inquiry and Word Count (LIWC); which have been extracted from one thousand magazine subscription reviews. Principal Component Analysis (PCA) is used as a dimensionality reduction technique. Further, along with five different variances of PCA and without PCA, scenarios were used to compare the performance of seven supervised machine learning techniques. It has been demonstrated that the Ensemble Bagged classifier with 3% PCA variance outperforms other six supervised methods resulting in 88% prediction accuracy.

Keywords: Deceptive review · Classification · LIWC · PCA · Supervised machine learning

1 Introduction

In online shopping, many consumers prefer to read reviews in order to make the best decisions on the product before purchasing it. Thus, reading the reviews became a profoundly influencing factor which affected the consumers' buying behavior in online shopping. The researchers have revealed from their studies that the ratings and reviews have a high impact on consumers' online buying behavior and making decisions in purchasing [1]. Products which have positive reviews and high ratings give a good impression to the consumer regarding the product and also many negative reviews, and low ratings have high potential to take consumers away. Therefore, many online merchants tend to manage an excellent online reputation while enhancing the popularity of the product [2].

W. S. H. Suhaili et al. (Eds.): CIIS 2020, AISC 1321, pp. 97–105, 2021.
https://doi.org/10.1007/978-3-030-68133-3_10

Since many consumers highly consider fellow consumers' reviews and trust them as personal recommendations when purchasing a product, many fake reviews are made by an organization or by an individual to mislead the consumer for many reasons [3]. So, consumers have to spend a considerable amount of time to get the correct decision based on the reviews which might be misleading. Consequently, the need for an assistant to identify genuine and fake reviews has arisen.

Detecting fake reviews became a popular topic among researchers recently, since deceptive reviews are widely used [4]. Thus, for the past ten years, many research works were carried out related to identifying fake reviews using various methods [3, 5, 6]. As in the article [3] stated, introducing an alliance approach to existing regulatory compliance systems and practice will offer a better deal with the problem. Moreover, they have mentioned that such an approach will not solve the problem but will certainly control it. Another research was conducted to identify and highlight the relevance of contextual indicators apart from textual which would assist internet users to detect possible deceptive reviews [6]. Moreover, computer science-based methods were introduced applying upcoming trends such as text mining, machine learning, deep learning, natural language processing and knowledge-based learning (using ontological features) [7]. Considering the machine learning methods, researchers have adapted approaches including supervised learning, unsupervised learning and semi-supervised learning [8].

Also, mainly two types of features namely content-based detection and behavior-based detection are used to identify the fake reviews based on three different types of characteristics identified as review centric attributes, product-centric attributes and reviewer centric attributes [9]. Data collection is one of the crucial tasks in a study related to fake review identification, in particular for machine learning methods. In the existing literature, four different categories of data set were formed depending on the construction methods; a ruled-based, human-based, filtering algorithm based and Amazon Mechanical Turk (AMT) based methods [7].

Another critical aspect of these studies which were done based on machine learning is feature engineering. The existing features can be divided into two categories which are text features and behavior features. Behavior features are the statistical characteristics of a particular reviewer's review behavior and the review. Under behavior features: maximum content similarity, maximum number of reviews, burst characteristics, percentage of the first review, repeated reviews, extreme rating behavior, percentage of positive reviews, reviewer deviation, characteristics of early reviews and review length were considered [7]. Under text features: Part of Speech (POS) [10], Bag of Words (BoW) [11] Linguistic Inquiry and Word Count (LIWC) [12, 13] and metadata features [14, 15] have been used. Under behavior features, statistical characteristics of a particular user's review behavior and his/her review are considered [7, 16]. As the classification methods under supervised machine learning; classifiers such as Support Vector Machine (SVM) [12, 13, 17] Logistic Regression (LR) [14, 15], Naive Bayes (NB) [13, 14, 18, 19], K-Nearest Neighbor (KNN) [18], Ensemble approaches [10], Decision Tree (DT) [8], Random Forest (RF) [20] and various types of Neural Networks [7] are adapted.

Among different approaches on detecting fake reviews, noticeably, the majority of the studies have been done using supervised learning [7]. Thus, a need for comparison of different classifiers based on different approaches in supervised learning has arisen.

In this study, the content-based detection was adapted with an amazon magazine subscriptions review dataset. Objectives of this study are to compare the prediction accuracy of several supervised learning classifiers with and without PCA for LIWC features and determine the best PCA variance of a robust classifier.

2 Methodology

Figure 1 elaborates the general flow adapted in this paper to identify the deceptive reviews using supervised machine learning techniques. It has main stages such as data collection, data preprocessing, feature extraction and training of clarifiers followed by the testing of the models to predict the classes. As an additional stage dimensionality reduction technique is adopted before feeding the features to the classifiers. Therefore, the proposed methodology will produce two results; one is with dimensionality reduction, and the other one is without dimensionality reduction.

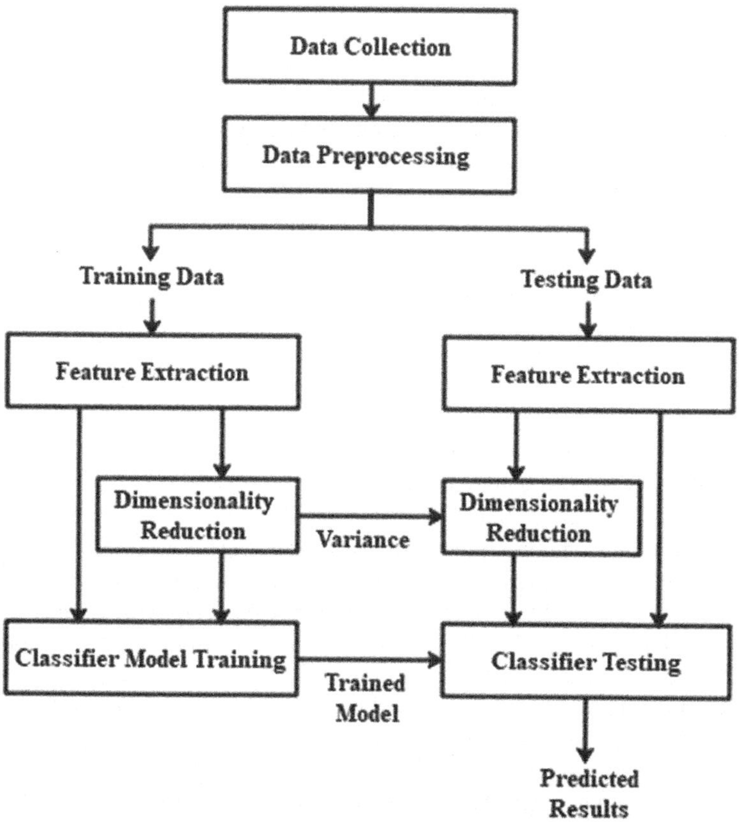

Fig. 1. The methodology adopted for deceptive review identification

2.1 Data Collection and Preprocessing

Data collected from the open-source online data repository called "Amazon Review Data (2018)" [21]. Only the Magazine Subscriptions review data were selected for this work. Data were already labelled either as true (genuine) or false (deceptive). Reviews from 2008 to 2018 were filtered in order to obtain the most recent behavior of reviewing. Then the duplicates, as well as data with missing values, were eliminated from the database. Finally, data belonging to both classes are separated as training and testing data set for further processing with 80:20 ratio.

2.2 Feature Extraction

This work focuses only on review text data; other ratings, helpfulness votes, and meta-data such as descriptions, category information, price, brand, and image features were ignored. Thus, the review centric features, LIWC is extracted from both training and testing data sets.

LIWC text analysis programme has been operated on the preprocessed data in order to analyze and extract features related to genuine and deceptive reviews. The LIWC application relies on an internal default dictionary that defines which words should be counted in the target text files. LIWC reads each designated text file, one target word at a time. As each target word is processed, the dictionary file is searched, looking for a dictionary match with the current target word. If the target word matches the dictionary word, the appropriate word category scale (or scales) for that word is incremented. As the target text file is being processed, counts for various structural composition elements (e.g., word count and sentence punctuation) are also incremented [22].

LIWC is a software which has been used as a text analyzing tool, specifically to analyze deceptions [7, 23]. It is a known factor that this tool can be used to analyze, extract and calculate significant features from a given set of text. In the latest release of the LIWC (2015), they have introduced four summary variables with 100 points scale from zero to hundred (0–100). Also it counts and groups the number of instances of nearly 6400 keywords, into approximately 90 output variables [22, 24]. This output dimension consists of word count and four summary variables:

- Analytical thinking: a high number reflects formal, logical, and hierarchical thinking; lower numbers reflect more informal, personal, here-and-now, and narrative thinking.
- Clout: a high number suggests that the author is speaking from the perspective of high expertise and is confident; low Clout numbers suggest a more tentative, humble, even anxious style.
- Authentic: higher numbers are associated with a more honest, personal, and disclosing text; lower numbers suggest a more guarded, distanced form of discourse.
- Emotional tone: a high number is associated with a more positive, upbeat style; a low number reveals greater anxiety, sadness, or hostility. A number around 50 suggests either a lack of emotionality or different levels of ambivalence.

In this study, all four summary variables have been adapted to measure the deceptiveness on the review.

2.3 Dimensionality Reduction

Despite feature extraction in order to remove the redundant and irrelevant properties and get significant features, dimensionality reduction techniques can be used. In this work, the most popular technique called PCA is used [25, 26]. Large datasets are increasingly common and are often difficult to interpret. PCA is a way for reducing the dimensionality of such datasets, increasing interpretability but at the identical time minimizing information loss. It does so by creating new uncorrelated variables that successively maximize variance.

Assume that the feature matrix X, after extraction is represented as m $\times$ n matrix; where n represents the samples and m represents the extracted features. PCA will linearly transform this matrix, X into another matrix, Y, also of dimension m $\times$ n, into m $\times$ m dimension matrix.

$$Y = PX \tag{1}$$

The Eq. 1 represents the change of basis. The rows of P to be the row vectors $p_1, p_2, \ldots, p_m$, and the columns of X to be the column vectors $x_1, x_2, \ldots, x_n$, then it can be interpreted as follows,

$$PX = (Px_1 Px_2 \ldots . Px_n) \tag{2}$$

Where original data X is projected in is being projected onto the columns of P, hence the rows of P, $\{P_1, P_2 \ldots P_m\}$ is a new basis for representing the columns of X; rows of P will be the PCA directions [27].

2.4 Classification

Supervised machine learning techniques Support Vector Machine (SVM), Linear Discriminant Analysis (LDA), K-Nearest Neighbour (KNN), Ensemble-Bagged (EB), Logistic Regression (LR) and Decision Tree (DT), Naive Bayes (NB) were trained and tested separately. The performances are compared with and without PCA classification.

Earlier ML with traditional statistical models were used without adapting any other methods for detecting fake reviews considering both behavior and content-based reviews identification. Among them, many types of research adapted with supervised learning classification methods since the results were more accurate than other methods in ML. Later researches more focused on hybrid methods to evaluate the performance of the supervised classification methods in the fake review identification field. Thus, PCA was adapted as feature reduction with supervised ML in hybrid methods [11].

The authors G.Vinodhini and RM. Chandrasekaran has shown that the adapting of feature reduction could affect the sentiment analysis of customer reviews. They used PCA on unigram features, and it increased the accuracy of SVM and Naïve Bayes classifiers [26]. Another research was conducted using active learning and PCA with three classifiers; Naïve Bayes (NB), Support Vector Machine (SVC) and Stochastic Gradient Descent (SGD) [11]. They used active learning methods for constructing labelled dataset for amazon product reviews and PCA for feature reduction. Then they evaluated the accuracy level on detecting fake reviews based on each classifier with and without

PCA. As a result, they have identified that the SGD classifier provides more accurate results, with 50% of feature reductions. Considering the existing work, evaluating the performance of supervised learning classifiers by adapting PCA has become more significant. Therefore, this work compares and finds the most robust supervised ML with appropriate PCA variance.

3 Results and Discussion

As described in Sect. 2.1 for each class, namely genuine and deceptive, five hundred (500) samples were selected (C = 2). Hence all together one thousand (1000) Magazine Subscriptions review data were used from the online data repository "Amazon Review Data (2018)" [21] and preprocessed. The number of training samples for each class is four hundred (400), and testing samples are one hundred (100). M = 93, which is the number of LIWC features extracted from all the samples. Each classifier was trained and tested for without PCA and five different PCA variance scenarios; where without PCA classification is mentioned as 0% PCA variance and other considered variance were 75%, 50%, 25%, 5% and 3%. Table 1 lists the prediction accuracy of each classifier for without PCA and other five PCA variance scenarios.

Table 1. Prediction accuracy of supervised machine learning techniques for selected various PCA variance

Supervised method	Accuracy Without PCA	PCA variance (%)				
		75	50	25	5	3
SVM	70	63	78.5	78.5	83.5	80
LDA	83	73.5	76.5	76.5	74.5	73.5
KNN	83	81.5	80.5	80.5	85	81.5
EB	88.5	85	81	82	86.5	88
LR	81.5	79.5	78.5	76.5	81	79.5
DT	82	83	76.5	76.5	81	83
NB	60	79	78.5	78.5	79	79

Overall, it can be seen that the EB supervised ML method has the highest accuracy as 88.5% among all the other methods without PCA variance. LDA and LR also show the highest prediction accuracy without PCA, while SVM and KNN have the highest prediction accuracy at the 5% PCA variance. DT resulted with 83% prediction accuracy at both 75%, and 3% PCA variances and NB shows highest prediction accuracy as 79% at 75%, 5% and 3% PCA variances. None of the methods were performed better with 50% and 25% PCA variances and also except DT and NB with 75% PCA variance prediction results of other methods were very poor.

Figure 2, compares the prediction accuracy of DT, NB, SVM, KNN, EB, LR and LDA classifiers for all five PCA variance and without PCA results. DT produces 82% accuracy without PCA, while NB performance is deficient as 60%. Gradually NB increased its performance up to 79% at 75% PCA variance and continued the same performance at each other PCA variance. Nevertheless, in sharp contrast to NB, DT has fluctuations in its performance and drastically drops its performance at 50% and 25% PCA variance. However, it started from a brief rise again in performance at 25% PCA variance and gradually increased the performance at 5% and 3% PCA variance. DT performs better with 83% accuracy while NB has 79% accuracy at 3% PCA variance. In both the cases, it is interesting to see that the prediction accuracy obtained with 75% variance and 3% variance are the individual highest and same.

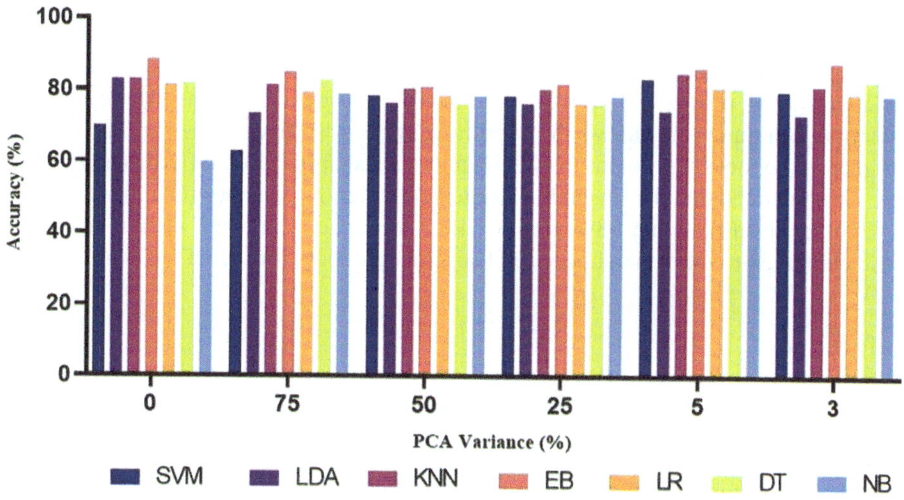

Fig. 2. Prediction accuracy of supervised classifiers with and without PCA variance

Performance accuracy of KNN is 83% without PCA while SVM has 70%. It shows a drop of performance at 75% PCA variance in SVM, but it increased performance continued until at 5% PCA variance then dropped again at 3% PCA variance to 80% accuracy. Compared to SVM, KNN shows better performance at each level and the performance accuracy is 81.5% at 3% PCA. Both SVM and KNN perform better at 5% PCA variance.

EB shows better performance for scenarios than LR and LDA. LDA performed better than LR without PCA, but it drastically dropped its performance in all other PCA variance. LR also performed low until at 25% PCA Variance. Then increased performance can be seen at 5% PCA variance, but dropped the accuracy again to 73.5% at 3% PCA variance. Same as in Fig. 1 it is captivating to observe that both EB and LR performance with low PCA variance 3% and 5% were very similar to the without PCA results.

Observing the results obtained it is evident that the Ensemble Bagged classifier outperforms other methods with an accuracy level reaching 88% with 3% PCA variance which it archives the objective of the study by determining EB as robust classifier with best PCA variance.

4 Conclusion

General deceptive review identification follows the standard processing steps of data collection preprocessing, feature extraction, and classification. In this paper, one thousand magazine subscription reviews, labelled as genuine and deceptive, were collected from an online database. Ninety-three (93) content-based features called Linguistic Inquiry and Word Count (LIWC), were extracted to form the feature matrix and Principal Component Analysis (PCA) is used to reduce the dimension as an additional step. Seven different supervised machine learning classifiers have been chosen for the comparative study. Further, the prediction accuracy results of each classifier were compared with and without PCA variance to find the most robust approach.

It has been shown that few classifiers perform better without PCA; but in general, adapting all the features may lead to overfitting of the model. Therefore, it is better to consider the best results with low PCA variance. The usage of PCA as a dimensionality reduction technique can address the overfitting issue. Also finding the appropriate PCA variance is the most significant in classification based applications. Furthermore, it has been demonstrated that the Ensemble Bagged classifier outperforms other methods with an accuracy level reaching 88% with 3% PCA variance. The result is significant, in a sense compared to other six supervised learning classifiers for the LIWC features.

References

1. State of Online Review (2019). http://learn.podium.com/rs/841-BRM-380/images/2017-SOOR-Infographic.jpg
2. Chen, R.Y., Guo, J.Y., Deng, X.L.: Detecting fake reviews of hype about restaurants by sentiment analysis (2014). https://doi.org/10.1007/978-3-319-11538-2_3
3. Malbon, J.: Taking fake online consumer reviews seriously. J. Consum. Policy **36**(2), 139–157 (2013). https://doi.org/10.1007/s10603-012-9216-7
4. Lau, R.Y.K., Liao, S.Y., Chi-Wai Kwok, R., Xu, K., Xia, Y., Li, Y.: Text mining and probabilistic language modeling for online review spam detection. ACM Trans. Manag. Inform. Syst. **2**(4), 1–30 (2011). https://doi.org/10.1145/2070710.2070716
5. Analyze and identify fake reviews and counterfeits. Fakespot.com (2019). https://www.fakespot.com/
6. Munzel, A.: Assisting consumers in detecting fake reviews: the role of identity information disclosure and consensus. J. Retail. Consum. Serv. **32**, 96–108 (2016). https://doi.org/10.1016/j.jretconser.2016.06.002
7. Ren, Y., Ji, D.: Learning to detect deceptive opinion spam: a survey. IEEE Access **7**(c), 42934–42945 (2019). https://doi.org/10.1109/access.2019.2908495
8. Rout, J.K., Singh, S., Jena, S.K., Bakshi, S.: Deceptive review detection using labeled and unlabeled data. Multimed. Tools Appl. **76**(3), 3187–3211 (2017). https://doi.org/10.1007/s11042-016-3819-y
9. Kumar, J.: Fake Review Detection Using Behavioral and Contextual Features (2020). http://arxiv.org/abs/2003.00807
10. Alsubari, S.N., Shelke, M.B.: Fake reviews identification based on deep computational linguistic features. Int. J. Adv. Sci. Technol. **29**(8), 3846–3856 (2020). https://www.scopus.com/inward/record.uri?eid=2-s2.0-85085685783&partnerID=40&md5=420584dbe09b455209097f3d3358fa57

11. Muhammad, F., Ahmed, S.: Fake review detection using principal component analysis and active learning. Int. J. Comput. Appl. **178**(48), 42–48 (2019). https://doi.org/10.5120/ijca2019919418
12. Ott, M., Choi, Y., Cardie, C., Hancock, J.T.: Finding deceptive opinion spam by any stretch of the imagination. In: ACL-HLT 2011 - Proceedings of the 49th Annual Meeting of the Association for Computational Linguistics: Human Language Technologies, vol. 1, pp. 309–319, July 2011. http://arxiv.org/abs/1107.4557
13. Olagunju, T., Oyebode, O., Orji, R.: Exploring key issues affecting african mobile ecommerce applications using sentiment and thematic analysis. IEEE Access **8**, 1–1 (2020). https://doi.org/10.1109/access.2020.3000093
14. Sihombing, A., Fong, A.C.M.: Fake review detection on yelp dataset using classification techniques in machine learning. In: Proceedings of the Fourth 2019 International Conference on Contemporary Computing and Informatics (IC3I) (2019)
15. Jindal, N., Liu, B.: Opinion-spam-and-analysis-WSDM-08.pdf. In: Proceeding of the First ACM International Conference on Web Search and Data Mining (2008)
16. Mukherjee, A., Liu, B., Glance, N.: Spotting fake reviewer groups in consumer reviews. In: WWW'12 - Proceedings of the 21st International Conference on World Wide Web, pp. 191–200 (2012). https://doi.org/10.1145/2187836.2187863
17. Ott, M., Cardie, C., Hancock, J.T.: Negative deceptive opinion spam. In: NAACL HLT 2013 - 2013 Conference of the North American Chapter of the Association for Computational Linguistics: Human Language Technologies. Proceedings of the Main Conference, pp. 497–501 (2013)
18. Hammad, A.A., El-Halees, A.: An approach for detecting spam in arabic opinion reviews. Int. Arab. J. Inf. Technol. **12**(1), 10–16 (2015)
19. Li, F., Huang, M., Yang, Y., Zhu, X.: Learning to identify review spam. IJCAI International Joint Conference on Artificial Intelligence, pp. 2488–2493 (2011). https://doi.org/10.5591/978-1-57735-516-8/ijcai11-414
20. Etaiwi, W., Naymat, G.: The impact of applying different preprocessing steps on review spam detection. Procedia Comput. Sci. **113**(September), 273–279 (2017). https://doi.org/10.1016/j.procs.2017.08.368
21. Amazon Review Data (2018). https://nijianmo.github.io/amazon/index.html
22. Pennebaker, J.W., Booth, R.J., Boyd, R.L., Francis, M.E.: Linguistic Inquiry and Word Count: LIWC2015 (2015)
23. Patel, N.A., Patel, R.: A survey on fake review detection using machine learning techniques. In: Proceedings - 2018 4th International Conference on Advances in Computing, Communication and Automation, ICACCA 2018 (2018)
24. Pennebaker, J.W., Boyd, R.L., Jordan, K., Blackburn, K.: The development and psychometric properties of LIWC2015. Austin, TX Univ. Texas Austin (2015). https://doi.org/10.15781/t29g6z
25. Wan, S., Tang, J., Yang, X.: Research on big data of customer product reviews based on text sentiment extraction and statistical analysis. Acad. J. Humanit. Soc. Sci. ISSN **3**(3), 21–25 (2020). https://doi.org/10.25236/AJHSS.2020.030303
26. Vinodhini, G., Chandrasekaran, R.M.: Opinion mining using principal component analysis based ensemble model for e-commerce application. CSI Trans. ICT **2**(3), 169–179 (2014). https://doi.org/10.1007/s40012-014-0055-3
27. Partridge, M., Jabri, M.: Robust principal component analysis. Neural Networks Signal Process. Proc. IEEE Work. **1**(May), 289–298 (2000). https://doi.org/10.1201/b20190-2

A Robust Ensemble Method for Classification in Imbalanced Datasets in the Presence of Noise

Chongomweru Halimu(✉) and Asem Kasem

School of Computing and Informatics, Universiti Teknologi Brunei, Gadong, Bandar Seri Begawan, Brunei

p20181008@student.utb.edu.bn, asem.kasem@utb.edu.bn

Abstract. Noise and class imbalance are two common data characteristics related to the quality and nature of many real-world data sources, which usually negatively affect the performance of many machine learning classification algorithms. A wide range of studies have investigated the problem of class imbalance and noise in isolation, and very few of them have studied their combined effect. In this paper, we propose a robust bagging-based ensemble method that tries to pay attention to both problems combined. The proposed method is based on the idea of Balanced Bagging (BB) to balance the bootstraps, but with a different sampling process, in which the probability of selecting an instance will be based on its level of hardness, i.e. the probability of an instance being misclassified irrespective of the choice of the classifier. The approach of the proposed method is based on estimating the hardness of each instance in a training dataset, and ensuring that bootstraps are balanced, and at the same time have instances of varying degrees of hardness (Easy, Normal, and Hard). We evaluate the performance of the proposed method on 30 synthetic imbalanced datasets with different levels of noise and imbalance ratios and compare its performance against the BB method. We observe that the proposed method performs significantly better than BB regardless of the noise level or imbalance ratio. Furthermore, we calculate the Equalized Loss of Accuracy (ELA) to assess the robustness of both methods under different levels of noise. The results indicate that the proposed method is more robust (not affected by noise as much) compared to BB. The Wilcoxon signed rank statistical test shows that there is a significant difference in both, performance and robustness, between the proposed method and BB, suggesting that representing varying levels of hardness in bootstraps is a better bootstrapping approach that improves the performance of ensemble methods.

Keywords: Class imbalance · Data noise · Instance hardness · Balanced bagging · Equalized loss of accuracy

1 Introduction

Class imbalance and noise are among the top data quality challenges in the machine learning community. Class imbalance problem happens when one class (majority class) has more instances than the opposite class (minority class). On the other hand, various

W. S. H. Suhaili et al. (Eds.): CIIS 2020, AISC 1321, pp. 106–118, 2021.
https://doi.org/10.1007/978-3-030-68133-3_11

definitions of noise have been presented in machine learning and statistics literature. The majority of them agree that noise presents inaccuracies that obscure the relationship between features of an object and its class [1]. The presence of class imbalance and noise in the training set, which is always unavoidable in the real-world domains, generally affects the classification performance of most learning algorithms.

When the dataset is imbalanced, traditional Machine Learning (ML) methods tend to bias towards the majority class, which results in a high classification accuracy but with very low detection rates towards the minority class. Furthermore, a dataset is assumed to be noisy if comprises erroneous data instances. The erroneous instances can appear in the input features (referred to as attribute noise) and the target labels (class noise). Zhu and Wu [2], distinguishes between attribute noise and class noise, the former modifies the observed values, whereas the latter mislabels the class in the observed labels allocated to an instance. Several studies in the literature have reported that class noise is more harmful to classification performance as compared to attribute noise [3]. Noise in a dataset can blur the decision boundary of a learning algorithm that separates the classes. It can cause a model to overfit in order to accommodate the erroneous data points. Noise effect on the classification performance of learning algorithms is determined by two issues; The Level of noise, which refers to the total volume of noise in the dataset, and Noise distribution that indicates the type of noise in the dataset. i.e. the proportion of erroneous instances that belong to the minority and majority class.

In the literature, whereas many works have studied the effects of class imbalance and noise on the performance of classification algorithms in isolation, very few of them have tried to address their combined effects, the majority of the combined studies concentrate on balanced problems and in traditional learning algorithms [4].

Ensemble learning, which incorporates data sampling techniques (under-sampling or oversampling) into bagging or boosting, has been highlighted as an effective way of addressing the problem of class imbalance. However, despite their highlighted improvement in classification performance, it has been reported that they cannot fully survive the effects of noise [5]. Since, majority of them tend to ignore the noisy instance during the bootstrap sampling process, where bootstraps instances are sampled based on uniform probability distribution with replacement. This creates the possibility of randomly generating a variety of bootstraps with a high number of noisy instances that might be hard to classify, which might eventually affect the overall classification performance of an ensemble. Nevertheless, different techniques have been proposed to filter and eliminate the noisy instance from the dataset as a means of solving the noise problem [3]. However, researchers have reported the undesired effects of these noise filtering techniques, such as discarding of instances in the boundaries between classes.

The concept of Instance Hardness (IH) presents another opportunity of dealing with erroneous instances. IH is a measure that specifies the degree of complexity in, correctly classifying a given instance in a dataset [6]. This implies that each instance in a dataset has a property that suggests its probability of being classified incorrectly regardless of the choice of the classifier.

In this study, we propose a robust bagging-based ensemble method that tries to pay attention to both problems of class imbalance and noise. The proposed method is based on the idea of Balanced Bagging (BB) of balancing the bootstraps, but with a completely different sampling process. Unlike BB that uses uniform probability distribution with replacement to sample instances into the bootstraps, the proposed method influences this process based on instances levels of hardness. To estimate the hardness of an instance, various classifiers are trained, then the hardness of an instance is estimated by the percentage of incorrect classifications made by a pre-selected set of classifiers, which are built from other instances in the dataset.

To evaluate the effectiveness of the proposed method, we carry out experiments on 30 synthetic imbalanced datasets (Table 1) with different levels of noise and compare its performance with Balanced Bagging in terms of AUC. Furthermore, we calculate the Equalized Loss of Accuracy (ELA) [7] to analyze the effect of noise in both methods as the noise level increases. Finally, we conduct statistical tests to ascertain if there exists any statistically significant difference in performance between the proposed method and BB in terms of AUC and ELA across all the datasets.

The rest of the paper is organized as follows; in Sect. 2 we present related work. We discuss the proposed method in Sect. 3, this is followed by experimental design in Sect. 4. We then present the results and discussions in Sect. 5 and finally conclude our study in Sect. 6.

2 Related Works

Class imbalance and noise are among the most common unavoidable data quality challenges affecting the majority of real-world classification problems [3, 8]. Numerous studies in the literature have investigated the problem of noise and class imbalance but in isolation. Limited consideration has been given to the challenge of learning from imbalanced noisy data. In this section, we present a summary of the existing work of class imbalance and noise that is related to our study.

Various methods for handling class imbalance problems have been proposed in the previous studies, but not from a perspective of noise. Japkowicz et al. [9] investigate the class imbalance problem highlighting particular aspects of class imbalance that are most difficult for learning. They went ahead to compare different techniques for addressing the class imbalance problem. In another study, Joffrey et al. [8] present a comprehensive survey on the classification of class imbalance problems. They mainly focused on data level and algorithm level techniques for addressing the class imbalance problem. Data level methods which include sampling-based techniques have received the most attention and the majority of class imbalance solutions have been derived from either under-sampling or oversampling techniques. For instance, Chawla et al. [10] proposed a Synthetic Minority Oversampling Technique (SMOTE) that tries to balance the class distribution of an imbalanced dataset by creating new synthetic instances into the minority class.

Furthermore, ensemble methods have been combined with sampling-based techniques to solve the challenge of class imbalance. Chawla et al. in their study [10], introduced a new method called SMOTEBoost for solving the class imbalance problem, their proposed method is based on SMOTE and boosting techniques. In another study [11], Liu et al. proposed an extension of the traditional Bagging ensemble method specialized for class imbalance problems called Balanced Bagging (BB).

The presence of noise in a dataset is mostly unavoidable in the real-world domain, it's now an important issue that researchers have to take into account when designing a learning algorithm [2]. Noise may appear in two forms: feature noise and class-label noise. Unless the former is very strong, the latter is considered the most harmful for learning and generalization [1]. Various techniques have been proposed in solving the problem of noise, including noise filtering techniques, noise cleaning, and robust classifiers [12]. A classifier is alleged to be robust to noise if its classification performance does not worsen significantly if faced with different levels of noise.

The majority of ensemble methods are considered to be robust learners as compared to single learners when dealing with the problem of noise. Sluban et al. [13], proposed an ensemble learning scheme for dealing with noise, where noisy instances are ranked based on the rate of inaccurate predictions performed by the classifier.

From the literature, we can observe that more research has been carried out on the problem of noise and class imbalance but in isolation. It is, therefore, important to consider both class imbalance and noise problems because they are pervasive in today's real-world domain and they move and in hand.

3 Proposed Method

The proposed method is based on Balanced Bagging (BB) [11]. BB is an ensemble method specialized for solving class imbalance problems. It is an extension of the traditional Bagging method, the idea behind BB is to under-sample instances from the majority class in each of the bootstraps in order to balance the majority class with the same number of instances as the minority class. Our motivation is to utilize the BB bootstrap aggregating process of balancing the bags but with a completely different sampling procedure. Unlike BB that uses uniform probability distribution with replacement to sample instances into the bootstraps, the proposed method influences this process based on instances levels of hardness (the probability of an instance being misclassified, irrespective of the choice of classifier).

We aim to influence the process of selecting bootstrap instances by first estimating the hardness of each instance (grouped into Easy, Normal, and Hard instances). To estimate the hardness of an instance, various classifiers are trained, then the hardness of an instance is estimated by the percentage of incorrect classifications made by a pre-selected set of classifiers, which are built from other instances in the dataset. We use k-Disagreeing Neighbors (kDN) as a measure for estimating instance hardness. kDN measure was introduced in [6] as a measure of instance hardness (IH). It is the fraction of k nearest neighbor of an instance that doesn't share its class label as shown in Eq. 1. Where $kDN(x)$ of an instance x, whose k nearest neighbors are denoted by $kDN(x)$ and $label(x)$ is the class label of example x.

$$kDN(x) = \frac{|x' | x' \in kNN(x) \wedge label(x') \neq label(x)|}{k}. \tag{1}$$

From the definition above, the hardness of an instance is represented by a real value between 0 and 1, where an instance with a hardness value close to 0 is more likely to be correctly classified, whereas the opposite is true for that instance with a hardness value close to 1. This implies that an instance with an estimated hardness value higher than a predefined threshold t_h is categorized as Hard (H), and with hardness value below the threshold t_e is categorized as Easy (E), or else as Normal (N). In this study, we set the default IH thresholds for Easy and Hard instances as $t_e = 0.33$ and $t_h = 0.66$, respectively, with an aim of maintaining equal ranges of hardness.

4 Experimental Design

Our experiments are organized in two stages, in stage one we carry out experiments on the imbalanced synthetic datasets with different Noise Nevels (NL), this will help us examine the influence of noise and class imbalance on the performance of the proposed method against Balanced Bagging.

We used a 5-fold cross-validation technique repeated 5 times, where each 5-fold cross-validation is calculated 5 different times with different random seeds. For fairness and uniformity, we used an ensemble size of 10 (n_estimators), and a Decision Tree algorithm with its default parameters as a base estimator for both methods being studied. For IH estimation, we used Logistic Regression (LR), Decision Tree (DT), k-Nearest Neighbor (kNN), and Gaussian Naive Bayes (NB), as a collection of both linear (LR) and nonlinear algorithms (kNN, DT, NB), and both parametric (NB, LR) and non-parametric algorithms (kNN, DT), as well as instance-based (kNN) and model-based (DT, NB, LR) ones. This enables a diverse and probably more reliable estimate of instance hardness.

To avoid data leakage from the test sets to the training phase, IH estimation is performed multiple times for each of the original imbalanced datasets, once in every fold of cross-validation, applied only to the training partition of the dataset.

In the second stage, we perform the Wilcoxon signed-rank statistical test [14] to statistically analyze and validate if there exists any significant difference in performance and robustness between the proposed method and BB in terms of AUC and ELA. The findings will enable us to draw more confident and reliable conclusions.

All the experiments were carried out in Python 3.7 using scikit-learn library, For statistical significance tests, we used KEEL's statistical analysis software [15].

4.1 Experimental Datasets

In this study, we used 30 synthetic binary imbalanced datasets with different levels of noise, i.e. disturbance ratio (the last number in dataset name) applied to create noisy examples in the dataset. They are publicly available on KEEL online repository [15]. Table 1 summarizes all their respective properties which include: Dataset id (ID), Total number of instances (#Inst), Percentage of Majority instances (%Maj), Percentage of minority instances (%Min), and Imbalance Ratio (IR). In this study, we consider IR as the ratio of percentages of Majority to Minority instances, i.e. $IR = \frac{\%Maj}{\%Min}$.

4.2 Evaluation Metrics

Table 1. Synthetic Imbalanced Datasets with different levels of Noise

ID	Dataset	#Feat	#Inst	%Maj	%Min	IR
S1	03subcl5-600-5-0	2	600	83.3	16.7	5
S2	03subcl5-600-5-30	2	600	83.3	16.7	5
S3	03subcl5-600-5-50	2	600	83.3	16.7	5
S4	03subcl5-600-5-60	2	600	83.3	16.7	5
S5	03subcl5-600-5-70	2	600	83.3	16.7	5
S6	03subcl5-800-7-0	2	800	87.5	12.5	7
S7	03subcl5-800-7-30	2	800	87.5	12.5	7
S8	03subcl5-800-7-50	2	800	87.5	12.5	7
S9	03subcl5-800-7-60	2	800	87.5	12.5	7
S10	03subcl5-800-7-70	2	800	87.5	12.5	7
S11	04clover5z-600-5-0	2	600	83.3	16.7	5
S12	04clover5z-600-5-30	2	600	83.3	16.7	5
S13	04clover5z-600-5-50	2	600	83.3	16.7	5
S14	04clover5z-600-5-60	2	600	83.3	16.7	5
S15	04clover5z-600-5-70	2	600	83.3	16.7	5
S16	04clover5z-800-7-0	2	800	87.5	12.5	7
S17	04clover5z-800-7-30	2	800	87.5	12.5	7
S18	04clover5z-800-7-50	2	800	87.5	12.5	7
S19	04clover5z-800-7-60	2	800	87.5	12.5	7
S20	04clover5z-800-7-70	2	800	87.5	12.5	7
S21	paw02a-600-5-0	2	600	83.3	16.7	5
S22	paw02a-600-5-30	2	600	83.3	16.7	5
S23	paw02a-600-5-50	2	600	83.3	16.7	5
S24	paw02a-600-5-60	2	600	83.3	16.7	5
S25	paw02a-600-5-70	2	600	83.3	16.7	5
S26	paw02a-800-7-0	2	800	87.5	12.5	7
S27	paw02a-800-7-30	2	800	87.5	12.5	7
S28	paw02a-800-7-50	2	800	87.5	12.5	7
S29	paw02a-800-7-60	2	800	87.5	12.5	7
S30	paw02a-800-7-70	2	800	87.5	12.5	7

For evaluating the performance of the methods in this study, we use the Area Under Curve (AUC) evaluation metric. AUC has been recommended in the literature as a reliable metric for the evaluation of binary classifications with class imbalance problems [16]. On the other hand, we use the Equalized Level of Accuracy (ELA) [7], to measure the robustness of the proposed method against BB when faced with different levels of noise. ELA is a robustness metric proposed for evaluating the behavior of an algorithm with noisy data.

5 Results and Discussions

In this section, we present the AUC and Equalized Loss of Accuracy (ELA) scores obtained from the series of conducted experiments. In the discussion, we group the analyses by both, noise levels and datasets, to make it simpler to understand how the proposed method performs in different situations.

Table 2 groups the AUC and ELA results as per the Noise Level (NL) across multiple datasets. i.e. NL0 represents all the datasets that have 0 noise and NL70 representing all the datasets with the highest level of noise. On the other hand, Table 3 groups the

Table 2. AUC and equalized loss of accuracy (ELA) score for BB against the proposed method (BB_IH) for each noise level on multiple datasets

NL0	**AUC**		**ELA**	
Dataset	**BB**	**BB_IH**	**BB**	**BB_IH**
S01	0.946	**0.947**	***0.057***	0.056
S06	0.948	**0.96**	***0.055***	0.042
S11	0.916	**0.931**	***0.092***	0.074
S16	0.908	**0.918**	***0.101***	0.089
S21	0.93	**0.94**	***0.075***	0.064
S26	0.94	**0.941**	***0.064***	0.063

NL60	**AUC**		**ELA**	
Dataset	**BB**	**BB_IH**	**BB**	**BB_IH**
S04	0.792	**0.803**	***0.220***	0.208
S09	**0.783**	0.777	0.229	***0.232***
S14	0.802	**0.807**	***0.216***	0.207
S19	**0.823**	0.813	0.195	***0.204***
S24	0.804	**0.815**	***0.211***	0.197
S29	0.802	**0.812**	***0.211***	0.200

NL30	**AUC**		**ELA**	
Dataset	**BB**	**BB_IH**	**BB**	**BB_IH**
S02	0.82	**0.831**	***0.190***	0.178
S07	0.828	**0.836**	***0.181***	0.171
S12	0.842	**0.86**	***0.172***	0.150
S17	0.841	**0.844**	***0.175***	0.170
S22	0.834	**0.853**	***0.178***	0.156
S27	**0.861**	0.833	0.148	***0.177***

NL70	**AUC**		**ELA**	
Dataset	**BB**	**BB_IH**	**BB**	**BB_IH**
S05	**0.783**	0.777	0.229	***0.235***
S10	0.766	**0.767**	***0.247***	0.243
S15	0.791	**0.804**	***0.228***	0.211
S20	**0.79**	0.779	0.231	***0.241***
S25	0.827	**0.84**	***0.186***	0.170
S30	0.828	**0.83**	***0.183***	0.181

NL50	**AUC**		**ELA**	
Dataset	**BB**	**BB_IH**	**BB**	**BB_IH**
S03	0.779	**0.787**	***0.234***	0.225
S08	**0.791**	0.786	0.220	***0.223***
S13	0.823	**0.832**	***0.193***	0.180
S18	0.816	**0.834**	***0.203***	0.181
S23	0.848	**0.852**	***0.163***	0.157
S28	0.822	**0.841**	***0.189***	0.169

AUC and ELA results of each dataset perturbed with different noise levels. i.e. Dataset *03subcl5-600-5* with NL0 indicates 0 noise in that dataset whereas NL70 implies that the same dataset is perturbed with the highest level of noise.

Table 3. AUC and equalized loss of accuracy (ELA) score for the proposed method (BB_IH) against BB for each dataset on different noise levels.

03subcl5-600-5

	AUC		ELA	
NL	BB	BB_IH	BB	BB_IH
NL0	0.946	**0.947**	***0.057***	0.056
NL30	0.82	**0.831**	***0.220***	0.203
NL50	0.779	**0.787**	***0.284***	0.271
NL60	0.792	**0.803**	***0.263***	0.245
NL70	**0.783**	0.777	0.277	***0.287***

03subcl5-800-7

	AUC		ELA	
NL	BB	BB_IH	BB	BB_IH
NL0	0.948	**0.96**	***0.055***	0.042
NL30	0.828	**0.836**	***0.210***	0.197
NL50	**0.791**	0.786	0.268	***0.272***
NL60	**0.783**	0.777	0.274	***0.278***
NL70	0.766	**0.767**	0.299	***0.300***

04clover5z-600-5

	AUC		ELA	
NL	BB	BB_IH	BB	BB_IH
NL0	0.916	**0.931**	***0.089***	0.073
NL30	0.842	**0.86**	***0.193***	0.168
NL50	0.823	**0.832**	***0.227***	0.213
NL60	0.802	**0.807**	***0.250***	0.240
NL70	0.791	**0.804**	***0.267***	0.252

04clover5z-800-7

	AUC		ELA	
NL	BB	BB_IH	BB	BB_IH
NL0	0.908	**0.918**	***0.097***	0.087
NL30	0.841	**0.844**	***0.194***	0.188
NL50	0.816	**0.834**	***0.236***	0.211
NL60	**0.823**	0.813	0.223	***0.233***
NL70	**0.79**	0.779	0.268	***0.284***

paw02a-600-5

	AUC		ELA	
NL	BB	BB_IH	BB	BB_IH
NL0	0.93	**0.94**	***0.074***	0.063
NL30	0.834	**0.853**	***0.202***	0.177
NL50	0.848	**0.852**	***0.195***	0.188
NL60	0.804	**0.815**	***0.247***	0.230
NL70	0.827	**0.84**	***0.221***	0.206

paw02a-800-7

	AUC		ELA	
NL	BB	BB_IH	BB	BB_IH
NL0	0.94	**0.941**	***0.063***	0.062
NL30	**0.861**	0.833	0.170	***0.201***
NL50	0.822	**0.841**	***0.228***	0.202
NL60	0.802	**0.812**	***0.250***	0.234
NL70	0.828	**0.83**	***0.220***	0.219

For all the results in both Tables 2 and 3, the highest AUC score for each method is bolded and this implies better performance. Whereas the highest ELA score is bolded with italic and this indicates high loss hence poor robustness.

From Table 2, we observe a good performance of the proposed method across the majority of the datasets regardless of the noise level. For example, at NL0, the proposed method performed better than BB across all the datasets. However, as noise levels start to increase, we notice a slight variation in performance, where the AUC of both methods starts to drop. (this is well represented in Table 2). For instance, if you compare the results from NL0 and NL70, in NL0 all the AUC scores for both methods were above 90%, the highest score being 96% registered by the proposed method on the S06 dataset. Whereas in NL70 the performance across all methods did not go above 84%. Despite the drop in performance as noise levels increase, we are glad to report that the proposed method still maintained better performance across 26 out of 30 datasets regardless of the noise level. This perhaps implies that the modification that we proposed for sampling bootstraps based on their level of hardness resulted in a stable performance gain. We also observe a low ELA for the proposed method across the majority of the datasets regardless of the noise level as compared to BB which registered high ELA across many datasets. This further implies that the proposed method is robust (not much affected by noise) as compared to the BB.

From the results in Table 3, we observe a decrease in AUC performance for both methods as the noise levels increase. This was highly expected since the quality of the data becomes worse every time the noise level increases.

Furthermore, we notice an interesting pattern across the majority of the datasets; ELA increases with an increase in noise level as shown in Fig. 1. where we plot ELA against noise levels. Interestingly, in one dataset (03subcl5-800-7) with a high imbalance ratio of 7, at NL70 we observe a high ELA for the proposed method, yet it performed better than BB in terms of AUC. We cannot clearly explain this situation, perhaps this

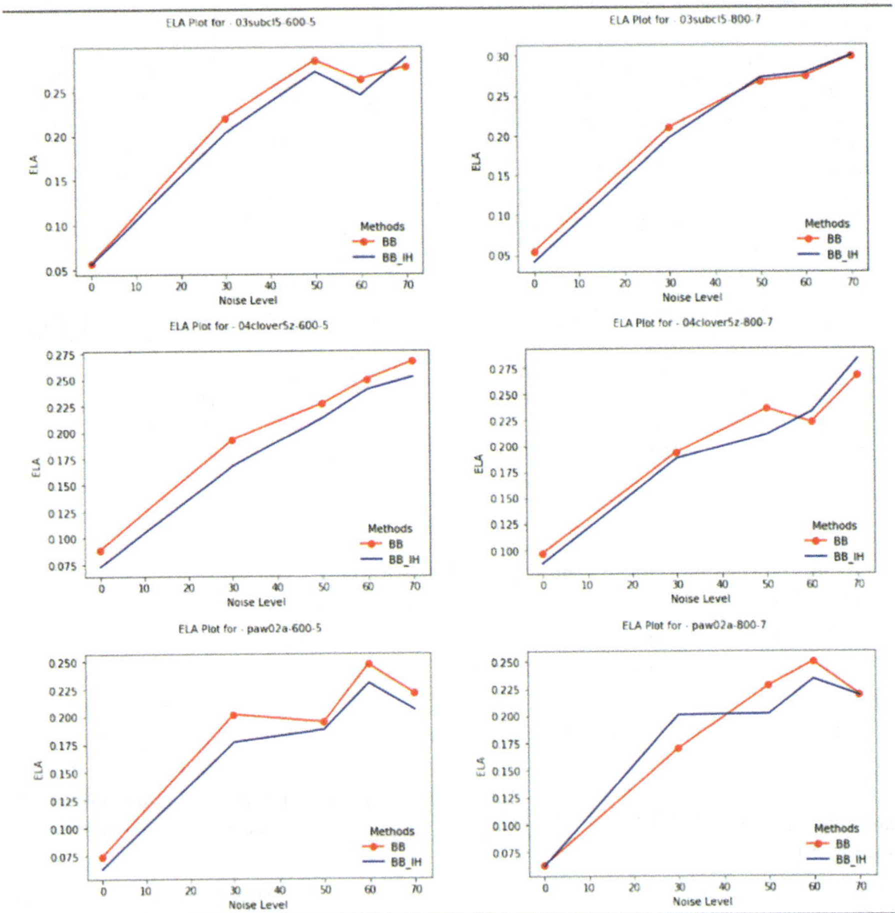

Fig. 1. Equalized Loss of Accuracy (ELA) Scores Vs NL for the proposed method (BB_IH) against BB for each Dataset.

simply implies that, even though an algorithm might perform better than the other in terms of AUC, that alone does not qualify it to be a more robust algorithm than the other. But more experiments in the future are needed to verify this. For our case, the proposed method registered a low ELA across the majority of the datasets regardless of the noise level which qualifies it to be a more robust method in handling noise as compared to BB.

5.1 Statistical Tests and Results

We carried statistical tests to compare the proposed method against BB to ascertain if there exists a significant difference between the two methods. We used Wilcoxon paired signed test as per the recommendation made by [14], where they presented the right methodology of using the nonparametric statistical tests in comparing multiple algorithms across multiple datasets. In Table 4 we present the outcome of the Wilcoxon test applied on the AUC and ELA results at a significant level of 0.05.

Table 4. Results obtained by the Wilcoxon test on AUC and ELA scores

	AUC			
BB_IH vs	R^+	R^-	P-value	Hypothesis
BB	372.5	92.5	***0.003105***	H_0 rejected
	ELA			
BB vs				
BB_IH	379.0	86.0	***0.001864***	H_0 rejected

From the results in Table 4, we observe an overall performance of the proposed method (in terms of AUC) obtaining higher ranks and rejecting the null hypothesis (H_0) of the equivalence with a p-value of ***0.003105.*** On the other hand, BB achieved higher ranks in terms of ELA and rejecting the Null Hypothesis of equivalence with a p-value of ***0.001864.*** This implies that BB was much affected by noise as compared to the proposed method. Suggesting that the proposed method is more robust than the BB.

5.2 Discussion of Results

The general conclusion obtained from the above results is that building an ensemble composed of balanced bootstraps containing instances with varying degrees of hardness will produce a consistent and noise-tolerant (robust) model that can better deal with the problem of class imbalance in the presence of noise. The intuitive explanation for this is that when a classifier is trained on a section of instances from a dataset with varying degrees of hardness, the classifier will have a better chance to learn the underlying pattern to classify the data while maintaining two properties: learn from hard instances as well as easy ones, and at the same time not overfit to those that are easy or hard. Junnal li et al. [17], concur with our intuition in their study where they proposed a noise-tolerant algorithm that simulates actual training by generating synthetic noisy labels to train their model such that it does not overfit specific noise and thus gives better performance.

Both BB and the proposed method use the same strategy of eliminating the problem of class imbalance, i.e. by balancing their bootstraps during the bootstrapping process. However, throughout the findings of this study, we observe a consistent and improved performance of the proposed method under different levels of noise as compared to BB. This variation in performance can only be attributed to the modification in the sampling process made by our proposed method. Unlike BB that uses uniform probability distribution with replacement to sample instances into the bootstraps, the proposed method influences this process based on instances levels of hardness. The intuition behind this modification in the bootstrapping process is that building a model with a mixture of instances with different levels of hardness may produce a noise-tolerant model. This perhaps justifies the observed consistency in the performance of the proposed method under different noise levels as observed by both AUC and ELA scores. This intuition is much supported by different studies in the literature. For example, Aditya et al. in their study [18], proved that any algorithm that is consistent for classification for noisy distribution will also be consistent on noise-free distribution. They further reported that it is possible to get an optimal classification and ranking error given enough noisy instances, without the need to clean them.

6 Conclusion and Future Works

In this study, we proposed an ensemble method that combines the idea of Balanced Bagging (BB) and data complexity measures to achieve a better classification performance in situations of class imbalance and noise. The proposed method leverages the kDN measure of IH to influence the bootstrap aggregating process of balanced bagging. This is to ensure that balanced bootstraps also contain instances with varying degrees of hardness, since several works in the literature have indicated that training an algorithm with a portion of hard instances might produce a noise-tolerant model that might be able to robustly learn the underlying patterns in training data.

We performed a series of experiments on 30 synthetic imbalanced datasets with different noise levels and compared the proposed method against BB. Our results show that the proposed method performs better than BB in terms of AUC. We also measured the robustness of both methods using the Equalized Measure of Accuracy (ELA) metric; the findings demonstrate that the proposed method is more robust to noise as compared to BB. All our findings were affirmed by a non-parametric Wilcoxon signed-rank statistical test which demonstrated a significant difference in performance between the proposed method and BB both in terms of AUC and ELA.

Future work may include further analyses to visualize and inspect the distribution of instances within the bootstraps, and to analyze the degrees of hardness and relations to decision boundaries within the balanced bags. Additionally, more experiments can be performed on real-world datasets injected with noise and to compare the proposed method with other state-of-the-art noise filtering techniques. The reason why we chose to only compare the proposed method against BB in this study was to focus on the idea of modifying the bootstrapping process of BB, which allowed us to draw meaningful conclusions before setting the pace for future works.

References

1. Krawczyk, B.: Learning from imbalanced data: open challenges and future directions. Prog. Artif. Intell. **5**(4), 221–232 (2016). https://doi.org/10.1007/s13748-016-0094-0
2. Zhu, X., Wu, X.: Class noise vs. attribute noise: a quantitative study. Artif. Intell. Rev. **22**, 177–210 (2004). https://doi.org/10.1007/s10462-004-0751-8
3. Nigam, N., Dutta, T., Gupta, H.P.: Impact of noisy labels in learning techniques: a survey. In: Kolhe, M.L., Tiwari, S., Trivedi, M.C., Mishra, K.K. (eds.) Advances in Data and Information Sciences. LNNS, vol. 94, pp. 403–411. Springer, Singapore (2020). https://doi.org/10.1007/978-981-15-0694-9_38
4. Sáez, J.A., Galar, M., Luengo, J., Herrera, F.: Tackling the problem of classification with noisy data using multiple classifier systems: analysis of the performance and robustness. Inf. Sci. (Ny) **247**, 1–20 (2013). https://doi.org/10.1016/j.ins.2013.06.002
5. Walmsley, F.N., Cavalcanti, G.D.C., Oliveira, D.V.R., Cruz, R.M.O., Sabourin, R.: An ensemble generation method based on instance hardness. In: Proceedings of International Joint Conference Neural Networks, July 2018 (2018). https://doi.org/10.1109/ijcnn.2018.8489269
6. Smith, M.R., Martinez, T., Giraud-Carrier, C.: An instance level analysis of data complexity. Mach. Learn. **95**(2), 225–256 (2013). https://doi.org/10.1007/s10994-013-5422-z
7. Sáez, J.A., Luengo, J., Herrera, F.: Evaluating the classifier behavior with noisy data considering performance and robustness: the equalized loss of accuracy measure. Neurocomputing **176**, 26–35 (2016). https://doi.org/10.1016/j.neucom.2014.11.086
8. Leevy, J.L., Khoshgoftaar, T.M., Bauder, R.A., Seliya, N.: A survey on addressing high-class imbalance in big data. J. Big Data **5**(1), 1–30 (2018). https://doi.org/10.1186/s40537-018-0151-6
9. Japkowicz, N., Stephen, S.: The Class imbalance problem: a systematic study. Intell. data Anal. **6**, 429–449 (2002). https://doi.org/10.3233/IDA-2002-6504
10. Chawla, N. V, Lazarevic, A., Hall, L.O., Bowyer, K.W.: SMOTEBoost: improving prediction of the minority class in boosting, pp. 107–119 (2003)
11. Martínez-García, J.M., Suárez-Araujo, C.P., Báez, P.G.: SNEOM: a sanger network based extended over-sampling method. application to imbalanced biomedical datasets. In: Huang, T., Zeng, Z., Li, C., Leung, C.S. (eds.) ICONIP 2012. LNCS, vol. 7666, pp. 584–592. Springer, Heidelberg (2012). https://doi.org/10.1007/978-3-642-34478-7_71
12. Koziarski, M., Woźniak, M., Krawczyk, B.: Combined cleaning and resampling algorithm for multi-class imbalanced data with label noise. Knowl.-Based Syst. **204**, 106223 (2020). https://doi.org/10.1016/j.knosys.2020.106223
13. Sluban, B., Gamberger, D., Lavrač, N.: Ensemble-based noise detection: noise ranking and visual performance evaluation. Data Min. Knowl. Disc. **28**(2), 265–303 (2012). https://doi.org/10.1007/s10618-012-0299-1
14. Demšar, J.: Statistical comparisons of classifiers over multiple data sets. J. Mach. Learn. Res. **7**, 1–30 (2006). https://doi.org/10.1016/j.jecp.2010.03.005
15. Alcalá-Fdez, J., Fernández, A., Luengo, J., Derrac, J., García, S., Sánchez, L., Herrera, F.: KEEL data-mining software tool: data set repository, integration of algorithms and experimental analysis framework. J. Mult. Log. Soft Comput. **17**, 255–287 (2011). https://doi.org/10.1007/s00500-008-0323-y
16. Halimu, C., Kasem, A., Newaz, S.H.S.: Empirical comparison of area under ROC curve (AUC) and Mathew correlation coefficient (MCC) for evaluating machine learning algorithms on imbalanced datasets for binary classification. In: ACM International Conference Proceeding Series, pp. 1–6 (2019). https://doi.org/10.1145/3310986.3311023

17. Li, J., Wong, Y., Zhao, Q., Kankanhalli, M.S.: Learning to learn from noisy labeled data. In: Proceedings IEEE Computer Social Conference Computing Visual Pattern Recognition, June 2019, pp. 5046–5054 (2019). https://doi.org/10.1109/cvpr.2019.00519
18. Menon, A.K., van Rooyen, B., Natarajan, N.: Learning from binary labels with instance-dependent noise. Mach. Learn. **107**(2), 1561–1595 (2018). https://doi.org/10.1007/s10994-018-5715-3

Extreme Learning Machine Based Model Improved with Adaptive Activation Functions

Tales Lima Fonseca(✉) and Leonardo Goliatt

Computational Modeling Program, Federal University of Juiz de Fora, Juiz de Fora, Brazil
tales.lima@ice.ufjf.br, leonardo.goliatt@ufjf.br

Abstract. This paper proposes an approach combining an Extreme Learning Machine based model with adaptive activation functions for regression problems. Research related to machine learning seeks to solve a given problem in the best possible way, generating increasingly accurate generalist predictive models. On the other hand, these models become more complex, requiring massive computational power to train and make inferences. The use of adaptive activation functions aims at increasing the models' predictive capacity allowing generating better models without increasing their complexity. We evaluated the models' performance in a benchmark of ten regression problems. The results showed that Extreme Learning Machine with adaptive activation functions, in general, outperforms standard functions with the same architecture.

Keywords: Adaptive activation functions · Extreme Learning Machine · Regression problems

1 Introduction

In the beginning of the research on Artificial Neural Network (ANN) the works commonly used the activation functions such as Binary, Linear, Sigmoid and Hyperbolic Tangents [9]. These functions, separately or in combination through different layers, were and still are capable of solving different machine learning problems with satisfactory quality. Besides, the use of the appropriate activation function makes a total difference in the process of creating and learning an ANN [1], and even studies on which activation function or set of functions are best suited to the problem [16,22].

However, many ANN's may require a large number of layers and neurons to perform the function/problem approximation, which leads to increasingly larger and computationally slower models, both in their training and prediction stages [3]. Besides, these models end up impacting, during their training and prediction, the environment [25,26], in addition to being unfeasible in straight edge

W. S. H. Suhaili et al. (Eds.): CIIS 2020, AISC 1321, pp. 119–128, 2021.
https://doi.org/10.1007/978-3-030-68133-3_12

computing. Thus, in the opposite direction to increase the number of neurons and the number of layers, work has emerged related to neurons' learning power.

In order to bring greater predictive power to an artificial neural network, some studies have started to study activation functions with the ability to adapt its shape during the training stage [19]. The first work to explore this property sought to adapt a polynomial activation function, obtaining better results than traditional functions [21]. After this study, new works related to the adaptive activation function emerged as Hyperbolic Tangent [4], Splines [2,30], PReLU [10], and Soft Exponential [7]. All of these works proved, through numerical experiments, to overcome functions that cannot adapt and presented a fast convergence.

With the confirmation of the gains of using adaptive activation functions, different applied studies have appeared, such as classification of electrocardiographic arrhythmias [28], inference of gene expression [18], classification of images generated by computed tomography [32] and approximation of smooth and discontinuous functions, as well as solutions of differential equations linear and non-linear partials [15].

All adjustable parameters of an ANN can be learned through an optimizer. The most common, based on the descending gradient, is Backpropagation [23]. Despite being the most common, it can present a slower convergence, needing in some cases to study other ways of finding the optimal parameters, as L-BFGS [31] and Evolution Strategies [27]. In another direction, to obtain useful models with fast learning, Extreme Learning Machines (ELM) networks emerged [12]. A combination of optimization methods is also common, such as the combination of Backpropagation with ELM [33], finding better parameters than each one individually.

This paper proposes a study of Adaptive Activation Functions in an ELM based model, trained the output layer using Back Propagation, for a set of 10 benchmark datasets. The main positive contributions of the study can be summarized as follows: (1) in the best the authors' knowledge, this is the first contribution implementing ELM with Adaptive Activation Functions; (2) this paper highlights two ways to use Adaptive Activation Functions, Adaptive Layer and Adaptive Neuron.

The results show that the adaptive approach of the functions, combined with ELM, produces results superior to those of a standard function with the same architecture. This work has the main contribution in using ELM with adaptive activation functions, something that has not been exploited until the present moment, according to the authors' knowledge.

2 Materials and Methods

2.1 Datasets

In this work, to compare the proposed techniques, ten datasets related to regression problems were used. We obtained the datasets through PMLB [20], a broad benchmark suite for machine learning evaluation and comparison. We randomly extracted 20% of the complete database for testing. Table 1 presents a summary of the databases used in the numerical experiments, containing their name, quantity used in training and testing and, the number of features.

Table 1. Summarization of the databases used in numerical experiments.

Dataset	Problem	Train instances	Test instances	Features
1	1027_ESL	390	98	4
2	1028_SWD	800	200	10
3	1029_LEV	800	200	4
4	1030_ERA	800	200	4
5	1089_USCrime	37	10	13
6	1096_FacultySalaries	40	10	4
7	294_satellite_image	5148	1287	36
8	4544_GeographicalOriginalofMusic	847	212	117
9	503_wind	5259	1315	14
10	573_cpu_act	6553	1639	21

2.2 Extreme Learning Machine (ELM)

Extreme Learning Machine (ELM) [13] is a particular case of feedforward artificial neural network, where the vast majority is composed of only one hidden layer. Compared with other types of prediction models, including ANN, the ELM has advantages if it is fast learning, good generalization ability, and convenience in terms of modeling [8].

In ELM, all hidden parameters are assigned randomly, including those of the activation functions, and the output parameters are obtained using the generalized inverse of the output matrix of the last hidden layer [14]. The ELM output function, with only one hidden layer, used in this article is given by [24]:

$$\hat{y}(\mathbf{x}) = \sum_{i=1}^{L} \beta_i G(\mathbf{w}_i, b_i, \mathbf{x}) \tag{1}$$

where $\hat{y}$ is the ELM prediction associated to the input vector $\mathbf{x}$, $\mathbf{w}_i$ is the weight vector that multiplies the input vector $\mathbf{x}$ to generate the input value of the ith-neuron in the hidden layer, b_i is the bias of the ith-neuron in the hidden layer, β_i are output weights that multiply the output of the ith-neuron in the hidden layer, $G(\cdot)$ is the nonlinear activation function, and L is the number of neurons in the hidden layer. The parameters $(\mathbf{w}, b)$ are randomly generated (normally distributed with zero mean and standard deviation equals to one), and weights β_i of the output layer are determined analytically.

In a traditional ELM, the output weight vector $[\beta_1, ..., \beta_L]$ can be determined by minimizing [11]

$$\min_{\boldsymbol{\beta} \in \mathbb{R}^L} (||\mathbf{H}\boldsymbol{\beta} - \mathbf{y}|| + C||\boldsymbol{\beta}||^2) \tag{2}$$

where $\mathbf{y}$ is the output data vector, $\mathbf{H}$ is the hidden layer output matrix

$$\mathbf{H} = \begin{bmatrix} G_1(\mathbf{w}_1, b_1, \mathbf{x}_1) & \cdots & G_L(\mathbf{w}_L, b_L, \mathbf{x}_1) \\ \vdots & \ddots & \vdots \\ G_1(\mathbf{w}_1, b_1, \mathbf{x}_N) & \cdots & G_L(\mathbf{w}_L, b_L, \mathbf{x}_N) \end{bmatrix} \text{ and } \mathbf{y} = \begin{bmatrix} y_1 \\ \vdots \\ y_N \end{bmatrix}$$

is the output data vector with N the number of data points. The optimal solution is given by

$$\boldsymbol{\beta} = (\mathbf{H}^{\mathbf{T}}\mathbf{H})^{-1}\mathbf{H}^{\mathbf{T}}\mathbf{y} = \mathbf{H}^{\dagger}\mathbf{y}$$

where $\mathbf{H}^{\dagger}$ is the pseudoinverse of $\mathbf{H}$.

In this study, to adapt the parameters present in the activation functions, we modified the process of obtaining the parameters $\boldsymbol{\beta}$. Instead of calculating the pseudoinverse of $\mathbf{H}$, we use the traditional Back Propagation. This proposal generates a hybrid between ANN and ELM, where the hidden parameters are still fixed and randomly generated, and the final parameters are obtained through the Back Propagation.

2.3 Adaptive Activation Functions

In a traditional artificial neural network, we have the standard activation functions, which do not change their shape during the training phase. Despite this, we have the weights and biases that adapt during training to minimize/maximize a specific performance metric. This process of adapting the parameters allows the models, whether neural networks or not, to learn an approximate function of the training data, thus allowing the model to generalize to new data.

Adaptive activation functions emerged to increase the learning capacity of a neural network, allowing the learning of some parameters, such as weights and biases, during the training phase. In this way, an activation function can adapt to shapes that make it possible to find better minima/maximums of the objective function than standard activation functions.

In this work we study the following four activation functions and their adaptive forms:

$$\text{PReLU}(x, \alpha = 10^{-2}) = \begin{cases} \alpha x & \text{if } x \leq 0 \\ x & \text{otherwise} \end{cases}$$

$$\text{Quadratic}(x, \alpha = 0.1, \beta = 10^{-6}, \gamma = 10^{-6}) = \alpha x^2 + \beta x + \gamma$$

$$\text{Sigmoid}(x, \alpha = 1, \beta = 1) = \frac{\alpha}{1 + exp(-\beta x)}$$

$$\text{Swish}(x, \alpha = 1, \beta = 1) = \frac{\alpha x}{1 + exp(-\beta x)}$$

We fix the α, β, and γ values for the standard activation function at the values shown above. For adaptive cases, these values are the initial values and may converge to different ones during the training stage.

We propose two ways to build adaptive architecture, one called Adaptive Layer and the other one, Adaptive Neuron. In the Adaptive Layer, neurons in the same layer shared all the activation functions' adaptive parameters. On the other hand, in the Adaptive Neuron, each internal neuron of a layer has its adaptive parameters, without sharing it between them.

Using the Adaptive Layer strategy, the number of parameters to be adapted is less than or equal when compared with the Adaptive Neuronio strategy. There is a trade-off between the number of parameters and complexity to find all expected values.

3 Computational Experiments and Discussion

The same number of neurons, to each dataset, was used to create the architectures in order to verify the capacity of the adaptive activation functions. This decision ensures that when comparing the adaptive types of functions, the architecture is the same for each problem in terms of numbers of neurons, varying only the function parameters' existence. Furthermore, we created ELM with only one hidden layer, containing several neurons equal to the floor of the number of features divided by two.

For each database, each activation function and each adaptive type (Normal, Layer and Neuron), we trained the network using Adam optimizer [17] for 500 epochs with a batch size of 32 and a learning rate of 10^{-4} that was adjusted every ten epochs using a multiplicative factor of 0.99. Furthermore, we carry out 30 independent training and evaluation runs, calculating for each of them the Root Mean Squared Error (RMSE). Table 2 presents a summary of the results, showing the mean and standard deviation of the RMSE metric of the independent executions. It is possible to observe that, in general, the adaptive functions have a lower mean than the standard function, indicating that they can better minimize the RMSE. On the other hand, it is possible to observe that the results are similar among the adaptive functions, perhaps indicating the architecture's limit.

In order to provide a deeper comparative analysis of the results, we performed a One-way Analysis of variance (ANOVA) [6] with a confidence interval of 95%. The null hypothesis that the means between each adaptive type are statistically significantly different from each other. The results showed that, in some cases, the test rejected the null hypothesis. With a 95% confidence interval, we used the Tukey test [29] to check where the difference between the means existed. Figure 1 shows a count, with a maximum value equal to the number of databases, of which adaptive types of function showed a significant mean difference according to the Tukey test.

Table 2. Mean and standard deviation of the RMSE metric for the different types of function and adaptive types. Values in bold represent the best result for the metric evaluated.

Dataset	Activation function	Normal	Layer	Neuron
1	PReLU	1.225 ± 0.154	1.137 ± 0.174	1.137 ± 0.174
	Quadratic	1.281 ± 0.070	$\mathbf{1.020} \pm 0.200$	$\mathbf{1.020} \pm 0.200$
	Sigmoid	1.251 ± 0.104	1.181 ± 0.147	1.181 ± 0.147
	Swish	1.217 ± 0.155	1.129 ± 0.175	1.129 ± 0.175
2	PReLU	0.773 ± 0.029	0.764 ± 0.037	0.763 ± 0.037
	Quadratic	0.805 ± 0.005	0.757 ± 0.038	$\mathbf{0.752} \pm 0.039$
	Sigmoid	0.796 ± 0.018	0.786 ± 0.033	0.781 ± 0.033
	Swish	0.765 ± 0.031	0.759 ± 0.035	0.760 ± 0.033
3	PReLU	0.920 ± 0.042	0.908 ± 0.059	0.908 ± 0.059
	Quadratic	0.937 ± 0.008	$\mathbf{0.896} \pm 0.067$	$\mathbf{0.896} \pm 0.067$
	Sigmoid	0.935 ± 0.026	0.923 ± 0.043	0.923 ± 0.043
	Swish	0.914 ± 0.047	0.907 ± 0.060	0.907 ± 0.060
4	PReLU	2.014 ± 0.053	1.991 ± 0.086	1.991 ± 0.086
	Quadratic	2.039 ± 0.018	$\mathbf{1.953} \pm 0.110$	$\mathbf{1.953} \pm 0.110$
	Sigmoid	2.030 ± 0.042	2.014 ± 0.064	2.014 ± 0.064
	Swish	2.004 ± 0.064	1.988 ± 0.085	1.988 ± 0.085
5	PReLU	37.748 ± 4.831	37.290 ± 4.602	37.163 ± 4.661
	Quadratic	36.850 ± 2.205	36.270 ± 1.881	$\mathbf{36.073} \pm 2.003$
	Sigmoid	37.909 ± 3.094	37.509 ± 2.685	37.452 ± 2.667
	Swish	37.032 ± 4.564	36.597 ± 4.321	36.546 ± 4.360
6	PReLU	4.876 ± 0.735	4.699 ± 0.629	4.699 ± 0.629
	Quadratic	4.809 ± 0.560	$\mathbf{4.543} \pm 0.367$	$\mathbf{4.543} \pm 0.367$
	Sigmoid	4.787 ± 0.736	4.609 ± 0.575	4.609 ± 0.575
	Swish	4.838 ± 0.694	4.651 ± 0.571	4.651 ± 0.571
7	PReLU	1.246 ± 0.123	1.167 ± 0.065	$\mathbf{1.083} \pm 0.042$
	Quadratic	1.872 ± 0.093	1.172 ± 0.039	1.092 ± 0.035
	Sigmoid	1.310 ± 0.055	1.151 ± 0.040	1.148 ± 0.037
	Swish	1.205 ± 0.085	1.170 ± 0.060	1.098 ± 0.042
8	PReLU	0.647 ± 0.026	$\mathbf{0.593} \pm 0.028$	0.608 ± 0.023
	Quadratic	0.858 ± 0.015	0.607 ± 0.035	0.594 ± 0.027
	Sigmoid	0.704 ± 0.030	0.633 ± 0.026	0.642 ± 0.025
	Swish	0.626 ± 0.025	0.600 ± 0.029	0.606 ± 0.027
9	PReLU	4.947 ± 0.545	4.421 ± 0.569	4.419 ± 0.567
	Quadratic	6.221 ± 0.285	4.419 ± 0.572	$\mathbf{4.404} \pm 0.571$
	Sigmoid	4.778 ± 0.660	4.499 ± 0.541	4.478 ± 0.553
	Swish	4.711 ± 0.500	4.420 ± 0.570	4.414 ± 0.568
10	PReLU	13.752 ± 1.262	12.803 ± 1.115	12.387 ± 1.133
	Quadratic	16.337 ± 0.378	12.729 ± 1.160	$\mathbf{12.299} \pm 1.146$
	Sigmoid	13.383 ± 1.012	12.497 ± 1.028	12.439 ± 1.021
	Swish	13.589 ± 1.177	12.847 ± 1.193	12.300 ± 1.091

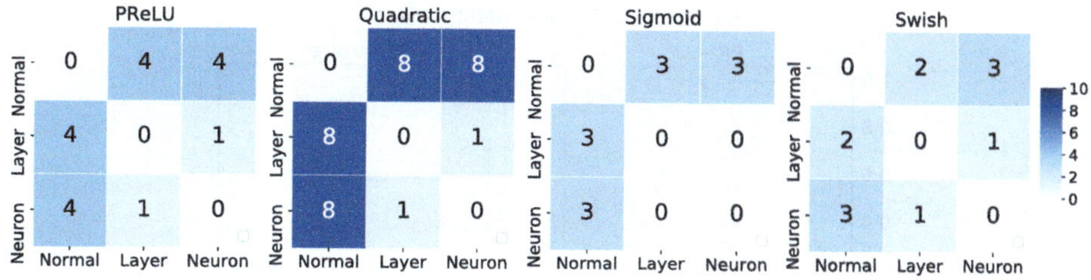

Fig. 1. Counting of adaptive function types that showed a significant mean difference according to the Tukey test.

The Adaptive Layer and the Adaptive Neuron, of three of the four functions, presented only one problem in which they had a significant difference, which indicates that in general, for these problems, the Adaptive Layer and the Adaptive Neuron approaches produced similar results. Notwithstanding this, it is possible to observe that both strategies present a considerable number of problems with a significant difference in the result, fostering the gain in using adaptive activation functions.

Finally, we use Performance Profiles [5] to facilitate the visualization and understanding of the results obtained in the experiments. Performance profiles are cumulative distributions of a performance metric. In this paper, RMSE is used to evaluate the performance of models. In general, a comparison is made between different algorithms to determine the one with the best performance on a specific set of problems. In general terms, the algorithm with the most significant area under the curve is considered the best among the analyzed group, considering the set of problems that were solved by them.

Figure 2 and Table 3 show the performance profiles' results for the different activation functions, different adaptive types, and the ten databases studied. It is possible to observe that the adaptive activation function approach is better for these databases than with the standard function. Besides, the Adaptive Neuron approach can perform better than the Adaptive Layer approach, despite presenting close results.

Table 3. Normalized area under performance profile curves for RMSE metric. Values in bold represent the best result for the metric evaluated.

Activation function	Normal	Layer	Neuron
PReLU	65.94%	92.98%	**98.95%**
Quadratic	76.13%	98.28%	**100.00%**
Sigmoid	70.63%	99.04%	**99.09%**
Swish	62.91%	91.45%	**99.23%**

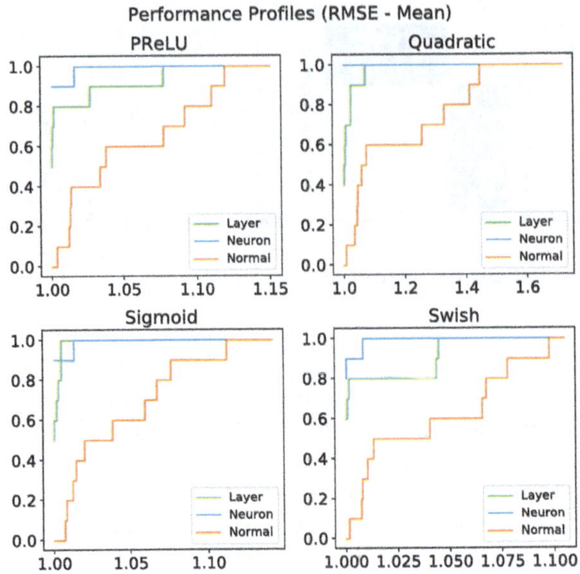

Fig. 2. Performance profile curves for RMSE Metric.

Although the research's preliminary results are encouraging, it is necessary to evaluate the proposed approach in a larger number of data sets. In this work, the analysis was limited to 10 data sets. In addition, the proposed strategy should also be tested on classification problems to assess better the contributions of the adaptive activation functions to solve these problems.

4 Conclusion

This paper investigates Adaptive Activation Functions in a modified version of the Extreme Learning Machine with Backpropagation to solve a regression benchmark's ten problems. The study showed that adaptive Layer and adaptive Neuron performed better than the standard function without increasing the architecture's size. This indicates that the search for an activation function's parameters increases its predictive power, making the need to search for increasingly larger architectures less priority. Regarding the activation functions, the Quadratic function was the one that obtained, in general, the best results for the analyzed databases and the one that presented a more significant performance gain in performing the learning of the parameters. This indicates that, with the analyzes carried out in this work, this adaptive activation function can result in better results in other problems. This study suggests limitations regarding the standard activation function and an alternative to increasing its predictive power. Further research involves searching for optimized architectures working together the adaptive activation functions to produce increasingly accurate solutions with a compact architecture size, reducing the model's complexity while improving its performance.

Acknowledgment. The authors acknowledge the financial support of CNPq (429639/2016-3), FAPEMIG (APQ-00334/18), and CAPES - Finance Code 001. The authors would like to thank Itaú Unibanco for hours released to its collaborator to develop this work.

References

1. Bishop, C.M., et al.: Neural Networks for Pattern Recognition. Oxford University Press, Oxford (1995)
2. Campolucci, P., Capperelli, F., Guarnieri, S., Piazza, F., Uncini, A.: Neural networks with adaptive spline activation function. In: Proceedings of 8th Mediterranean Electrotechnical Conference on Industrial Applications in Power Systems, Computer Science and Telecommunications (MELECON 96), vol. 3, pp. 1442–1445. IEEE (1996)
3. Canziani, A., Paszke, A., Culurciello, E.: An analysis of deep neural network models for practical applications. arXiv preprint arXiv:1605.07678 (2016)
4. Chen, C.T., Chang, W.D.: A feedforward neural network with function shape autotuning. Neural Netw. **9**(4), 627–641 (1996)
5. Dolan, E.D., Moré, J.J.: Benchmarking optimization software with performance profiles. Math. Program. **91**(2), 201–213 (2002)
6. Fisher, R.A.: Xv.—the correlation between relatives on the supposition of mendelian inheritance. Earth Environ. Sci. Trans. Roy. Soc. Edinb. **52**(2), 399–433 (1919)
7. Godfrey, L.B., Gashler, M.S.: A continuum among logarithmic, linear, and exponential functions, and its potential to improve generalization in neural networks. In: 2015 7th International Joint Conference on Knowledge Discovery, Knowledge Engineering and Knowledge Management (IC3K), vol. 1, pp. 481–486. IEEE (2015)
8. Guo, P., Cheng, W., Wang, Y.: Hybrid evolutionary algorithm with extreme machine learning fitness function evaluation for two-stage capacitated facility location problems. Expert Syst. Appl. **71**, 57–68 (2017)
9. Haykin, S.: Neural Networks - A Comprehensive Foundation, 2nd edn. Prentice Hall, Upper Saddle River (1998)
10. He, K., Zhang, X., Ren, S., Sun, J.: Delving deep into rectifiers: surpassing human-level performance on imagenet classification. In: Proceedings of the IEEE International Conference on Computer Vision, pp. 1026–1034 (2015)
11. Huang, G., Huang, G.B., Song, S., You, K.: Trends in extreme learning machines: a review. Neural Netw. **61**(Supplement C), 32–48 (2015)
12. Huang, G.B., Wang, D.H., Lan, Y.: Extreme learning machines: a survey. Int. J. Mach. Learn. Cybernet. **2**(2), 107–122 (2011). https://doi.org/10.1007/s13042-011-0019-y
13. Huang, G.B., Zhu, Q.Y., Siew, C.K.: Extreme learning machine: a new learning scheme of feedforward neural networks. In: 2004 IEEE International Joint Conference on Neural Networks. Proceedings, vol. 2, pp. 985–990. IEEE (2004)
14. Huang, G.B., Zhu, Q.Y., Siew, C.K.: Extreme learning machine: theory and applications. Neurocomputing **70**(1–3), 489–501 (2006)

15. Jagtap, A.D., Kawaguchi, K., Karniadakis, G.E.: Adaptive activation functions accelerate convergence in deep and physics-informed neural networks. J. Comput. Phys. **404**, 109136 (2020)
16. Karlik, B., Olgac, A.V.: Performance analysis of various activation functions in generalized MLP architectures of neural networks. Int. J. Artif. Intell. Expert Syst. **1**(4), 111–122 (2011)
17. Kingma, D.P., Ba, J.: Adam: a method for stochastic optimization. arXiv preprint arXiv:1412.6980 (2014)
18. Kunc, V., Kléma, J.: On transformative adaptive activation functions in neural networks for gene expression inference. bioRxiv p. 587287 (2019)
19. Lau, M.M., Lim, K.H.: Review of adaptive activation function in deep neural network. In: 2018 IEEE-EMBS Conference on Biomedical Engineering and Sciences (IECBES), pp. 686–690. IEEE (2018)
20. Olson, R.S., La Cava, W., Orzechowski, P., Urbanowicz, R.J., Moore, J.H.: PMLB: a large benchmark suite for machine learning evaluation and comparison. BioData Min. **10**(1), 36 (2017)
21. Piazza, F., Uncini, A., Zenobi, M.: Artificial neural networks with adaptive polynomial activation function (1992)
22. Ramachandran, P., Zoph, B., Le, Q.V.: Searching for activation functions. arXiv preprint arXiv:1710.05941 (2017)
23. Rumelhart, D.E., Hinton, G.E., Williams, R.J.: Learning representations by back-propagating errors. Nature **323**(6088), 533–536 (1986)
24. Saporetti, C.M., Duarte, G.R., Fonseca, T.L., da Fonseca, L.G., Pereira, E.: Extreme learning machine combined with a differential evolution algorithm for lithology identification. RITA **25**(4), 43–56 (2018)
25. Schwartz, R., Dodge, J., Smith, N.A., Etzioni, O.: Green AI. arXiv preprint arXiv:1907.10597 (2019)
26. Strubell, E., Ganesh, A., McCallum, A.: Energy and policy considerations for deep learning in NLP. arXiv preprint arXiv:1906.02243 (2019)
27. Sulistiyo, M.D., Dayawati, R.N., et al.: Evolution strategies for weight optimization of artificial neural network in time series prediction. In: 2013 International Conference on Robotics, Biomimetics, Intelligent Computational Systems, pp. 143–147. IEEE (2013)
28. Tezel, G., Özbay, Y.: A new neural network with adaptive activation function for classification of ECG arrhythmias. In: International Conference on Knowledge-Based and Intelligent Information and Engineering Systems, pp. 1–8. Springer (2007)
29. Tukey, J.W.: Exploratory Data Analysis, vol. 2, Reading, MA (1977)
30. Vecci, L., Campolucci, P., Piazza, F., Uncini, A.: Approximation capabilities of adaptive spline neural networks. In: Proceedings of International Conference on Neural Networks (ICNN 1997), vol. 1, pp. 260–265. IEEE (1997)
31. Wu, R., Huang, H., Qian, X., Huang, T.: A L-BFGS based learning algorithm for complex-valued feedforward neural networks. Neural Process. Lett. **47**(3), 1271–1284 (2018)
32. ZahediNasab, R., Mohseni, H.: Neuroevolutionary based convolutional neural network with adaptive activation functions. Neurocomputing **381**, 306–313 (2020)
33. Zou, W., Yao, F., Zhang, B., Guan, Z.: Back propagation convex extreme learning machine. In: Proceedings of ELM-2016, pp. 259–272. Springer (2018)

Rockfall Detection from Terrestrial LiDAR Point Clouds by Using DBSCAN with Clutter Removal Based on Grid Density

Pitisit Dillon[1,2](✉), Jessada Karnjana[1], and Pakinee Aimmanee[2]

[1] NECTEC, National Science and Technology Development Agency, 112 Thailand Science Park, Khlong Luang, Pathum Thani, Thailand
m6222040278@g.siit.tu.ac.th, jessada.karnjana@nectec.co.th

[2] Sirindhorn International Institute of Technology, Thammasat University, 131 Moo 5, Tiwanon Rd., Bangkadi, Muang, Pathum Thani, Thailand
pakinee@siit.tu.ac.th

Abstract. This paper proposes a simple method for rockfall detection from terrestrial LiDAR point clouds. The method consists of four steps: registration, subtraction, clutter removal, and spatial clustering. The paper contributes a straightforward method for clutter removal based on grid density, which is computational complexity inexpensive compared to the standard method based on nearest neighbor distance. Experimental results show that both are comparable in terms of identifying rockfall events. The proposed method can detect 21 events from 27 events from our simulations, and a conventional method can detect 23 events. The false-positive events of the proposed and conventional methods are 1 and 15, respectively. In contrast, for 52,000 points, the proposed method is about 16 times faster. Also, this paper suggests a simple means to estimate the parameters used in the spatial clustering algorithm.

Keywords: Rockfall detection · Terrestrial LiDAR point cloud · Clutter removal · DBSCAN

1 Introduction

A rockfall is a free fall movement of a detached segment of bedrock from a cliff or a very steep slope [4]. It can be considered as the faster type of landslide. Rockfalls cause damages not only to properties but also to lives. The former is mostly by impeded transportation and commerce due to blocked highways and waterways. The latter is as direct casualties from falling rocks [8]. The rockfall phenomena can be studied in a number of approaches, such as historical inventories, susceptibility assessment, frequency estimation, hazard assessment,

W. S. H. Suhaili et al. (Eds.): CIIS 2020, AISC 1321, pp. 129–139, 2021.
https://doi.org/10.1007/978-3-030-68133-3_13

and risk assessment [12]. Recently, a terrestrial laser scanner (TLS) has been used to study geological phenomena due to its advantage in acquiring high-resolution data [3,5,10,12]. For example, the TLS with the density-based spatial clustering of applications with noise (DBSCAN) has been applied in studying geomorphology of a rock glacier [10]. The similar technique has also been applied for rockfall detection [3,5,12]. A typical framework used to detect rockfalls from two point-cloud datasets, which are obtained from the TLS, consists of three parts: preprocessing, clutter removal, and spatial clustering [12], and a standard clutter removal algorithm is based on the nearest-neighbor-related approach, such as the nearest neighbor clutter removal (NNCR) [6]. However, the algorithm like NNCR has one problem concerning the computational time when the number of data points is large. Therefore, this paper aims to propose and discuss a comparable but more straightforward method for removing the clutter.

The rest of this paper is organized as follows. Section 2 provides background information used in the proposed method. Section 3 describes the detail of the proposed method. Section 4 reports experimental conditions and results. Discussion is made in Sect. 5, and Sect. 6 concludes this work.

2 Background

This section provides background information concerning an open-source software tool used to preprocess data points (i.e., CloudCompare) and a standard density-based spatial clustering method, called DBSCAN. Both are used together with our proposed clutter removal based on grid density.

2.1 CloudCompare

CloudCompare is an open-source 3D point cloud editing and processing software [2]. It was firstly designed to compare two point-cloud datasets directly. It can also perform many tasks, such as image registration, resampling, segmentation, and some statistical computation. One of its advantages is that CloudCompare can process more than 10 million data points on a standard laptop. In this work, CloudCompare is used to preprocess data in two steps: image registration and surface subtraction.

2.2 Density-Based Spatial Clustering of Applications with Noise (DBSCAN)

DBSCAN is one of the most common spatial clustering algorithms, which is based on data point density [7]. The algorithm collects points that are close to each other together (i.e., the points in high-density regions) and assigns the collection as a cluster. For points that lie in low-density areas or those not in the cluster, they are designated as noise. The DBSCAN concept and algorithm are briefly reviewed in this subsection for the completeness in itself.

Let p and q are points in a database D, and $d(p, q)$ denotes a distance between points p and q with regard to a distance function d, e.g., Euclidean distance or Manhattan distance. The *ε-neighborhood* of a point p, denoted by $N_\varepsilon(p)$, is defined as a set of points of which their distances to the point p is not greater than ε. That is, $N_\varepsilon(p) = \{q \in D \mid d(p,q) \leq \varepsilon\}$. A point p is said to be *directly density-reachable* for a point q if it satisfies two conditions. First, $p \in N_\varepsilon(q)$. Second, $N_\varepsilon(q)$ has at least `minPts` points, i.e., $|N_\varepsilon(q)| \geq$ `minPts`, where `minPts` is a positive integer. Note that the point q that satisfies the second condition is called a *core point* [7].

A point p is said to be *density-reachable* from a point q if there is a sequence of points $p_1, p_2, \ldots, p_n$, where $p_1 = q$, $p_n = p$, and p_{i+1} is *directly density-reachable* from p_i. A point p is said to be *density-connected* to a point q if there exists a point from which the point p and q are *density-reachable.*

Based on the concepts of density-reachability and density-connectivity, the DBSCAN algorithm groups together points in D and assigns them to members of a cluster C, which is a non-empty subset of D, if the following conditions are satisfied. First, for any points p and q, if $p \in C$ and q is *density-reachable* from p, then $q \in C$. Second, for all points p and all points q in C, p is *density-connected* to q. According to the definition of the cluster, the DBSCAN algorithm can discover a cluster by choosing a *core point* first and then retrieving all points that are *density-reachable* from that core point [7]. All other points that do not belong to any cluster are considered as noise. The abstract DBSCAN algorithm can be summarized in three steps [11]. First, find $N_\varepsilon(p)$ for all points p in D and identify core points. Second, join neighbor core points into clusters, ignoring all non-core points. Last, for each non-core point, add to a nearby cluster if the cluster is an *ε-neighborhood*; otherwise, assign it to noise.

3 Proposed Method

The proposed method is similar to the conventional rockfall detection method [12] based on DBSCAN with the nearest neighbor clutter removal (NNCR) [6] in the sense that it consists of the same four processes: surface registration, subtraction, clutter removal, and DBSCAN, as shown in Fig. 1 (left and middle). However, instead of deploying NNCR, which is computational time expensive, we propose a simple grid-density-based process to do the same task in the clutter removal.

The proposed method's inputs are two 3D point-cloud surfaces, and the proposed method can be summarized as follows. First, two point-cloud surfaces are registered before comparison so that both are geometrically aligned. Figure 1 (right) shows an example of two input point-cloud surfaces. Second, the difference between the two surfaces is determined. Note that the Cloud Compare software performs these first two steps. In order to determine the difference, a threshold value is set. The result of this subtraction is a set of data points of which their differences are above the threshold. Third, considered only in 2D, the subtraction result from the previous step, which is called a *surface-difference image*, is divided into equal square boxes of size w, as shown in Fig. 2 (b). Then, the number of data points in each box is counted. Let $c_{\mathtt{max}}$ and $c_{\mathtt{min}}$ are maximum and minimum numbers of data points in a box, respective. According to our observation, data points in any box with less than $(c_{\mathtt{max}} - c_{\mathtt{min}})/2$ data points can be considered clutters. All clutters are then removed from the surface-difference image. Fourth, the clutter-removed surface-difference image obtained from the previous step is clustered by DBSCAN to identify rockfall events.

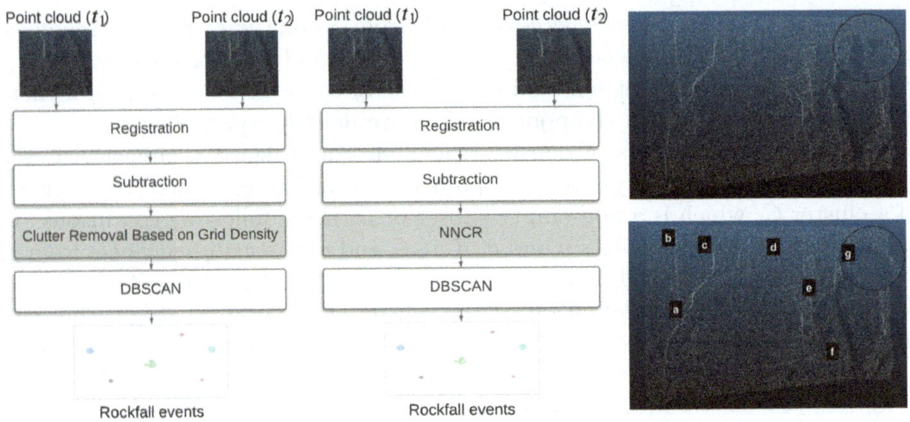

Fig. 1. Proposed method (left) and the conventional method (middle). Example of two input point-cloud surfaces: before rockfall events (right-top) and after the events (right-bottom). The rockfall events are marked by characters from a to g.

4 Experiments and Results

4.1 Data

Our experiments' point cloud dataset is taken from an open dataset provided by Abellan *et al.* [3,12] and available online [1]. The data points were acquired by a terrestrial laser scanner, Optech Intelligent Laser Ranging and Imaging System (ILRIS3D), from the Puigcercos cliff, Catalonia, Spain, as shown in Fig. 3. The wavelength of the infrared laser pulse of ILRIS3D is 1,535 nm. The distance d between the cliff's surface and the instrument is determined by the flight time Δt of the pulse, i.e., $d = c\Delta t/2$, where c is the speed of light [9]. The points' positions are initially recorded in the spherical coordinate system, i.e., (radial distance r, azimuthal angle θ, polar angle ϕ). Later, they are converted to coordinates in the Cartesian system (x, y, z) by the equation $(x, y, z) = (r \sin\theta \cos\phi, r \sin\theta \sin\phi, r \cos\theta)$. Not only the coordinate, but the instrument also records the reflective-light intensity for each data point as well. Thus, the standard format of each data point record can be represented by a quadruple (x, y, z, I), where I is the intensity. The maximum range of ILRIS3D is 700 m, with an error of 0.7 cm at 100 m.

In order to evaluate the performance of the proposed method, two datasets are required: one representing the surface before rockfall events and another representing the surface after the events. In this work, we generated the latter from the former by the following steps so that we can know the number and locations of the rockfall events exactly. First, P percent of the data points are randomly selected. Second, all selected points are randomly shifted in the $+x$-axis direction for a certain value (in cm) in an interval $[a, b]$. These two steps can represent natural noise in the measurement process and an error due to the instrument's installation, not at exactly the same locations, at two times. Third, a few simulated hemispherical holes with various diameters are added to the surface by modifying some y values of the data points. Some holes are generated by two to five partially-overlapped hemispherical holes. These holes can represent the deformation of the surface after the rockfall events. An example of the holes is shown in Fig. 1 (right). Note that in order to reduce the computational time in our simulation, 3D point-cloud inputs used in experiments are subareas or regions of the image in Fig. 3 (middle), as shown in Fig. 3 (bottom). The numbers of data points of all five regions are 15, 718, 41, 466, 33, 645, 13, 898, and 5, 385 points, for region numbers running from 1 to 5.

4.2 Experimental Conditions

The numbers of rockfall events in all five regions were as follows. Note that some events were generated from two to five hemispheres. These events are marked with the asterisk, as shown in Table 1. In region 1, there are 7 events from `a` to `g` with the largest radii of 25, 20, 7, 20, 60, 50, and 25 cm, respectively, and the smallest radius of `f` is 20 cm. In region 2, there are 9 events from `a` to `i` with the largest radii of 60, 40, 15, 40, 25, 40, 20, 30, and 60 cm, respectively, and the smallest radii of `a` and `d` are 25 and 20 cm, respectively. In region 3, there are 7 events from `a` to `g` with the largest radii of 7, 18, 10, 20, 50, 25, and 40 cm, respectively, and the smallest radius of `e` is 35 cm. In region 4, there are 7 events from `a` to `g` with the largest radii of 3, 25, 40, 40, 40, 25, and 20 cm, respectively, and the smallest radius of `e` is 10 cm. In region 5, there are 5 events from `a` to `d` with the largest radii of 4, 5, 14, and 25 cm, respectively, and the smallest radius of `d` is 18 cm.

The conditions used in our simulations are as follows. The parameter P, which is the percent of data points that are randomly selected for shifting, was set to 90%. The shifting interval $[a, b]$ was set to $[1, 2]$ because the resolution of ILRIS3D is around 2 cm. The box's size w was set to 10 cm because we assume that a rockfall event with a diameter less than 10 cm might not of importance in this work. The DBSCAN parameters used here are 10 and 5 for ε and `minPts`, respectively. Note that these parameters' values are suggested by the conventional method [12].

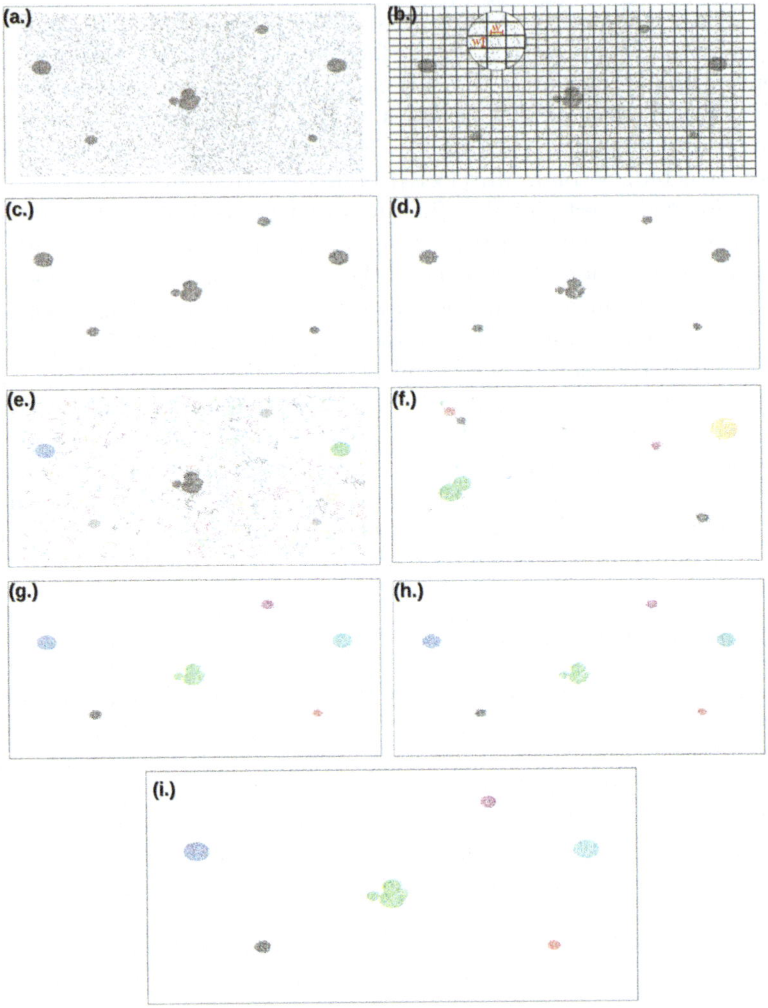

Fig. 2. Comparison of the results of the proposed method and the conventional method [12]: (a) surface-difference image from the subtraction process, (b) grid of equal square boxes of size w, (c) clutter-removed surface-difference image obtained from NNCR, (d) clutter-removed surface-different image obtained from the proposed grid-density-based method, (e) detected rockfall events when the surface-difference image is put as the input of DBSCAN (with parameters suggested by the conventional method) directly, (f) detected rockfall events when applying only DBSCAN with parameters estimated by the proposed grid-density-based method (without the clutter removal), (g) detected rockfall events by the conventional method, (h) detected rockfall events by the proposed method, and (i) ground truth.

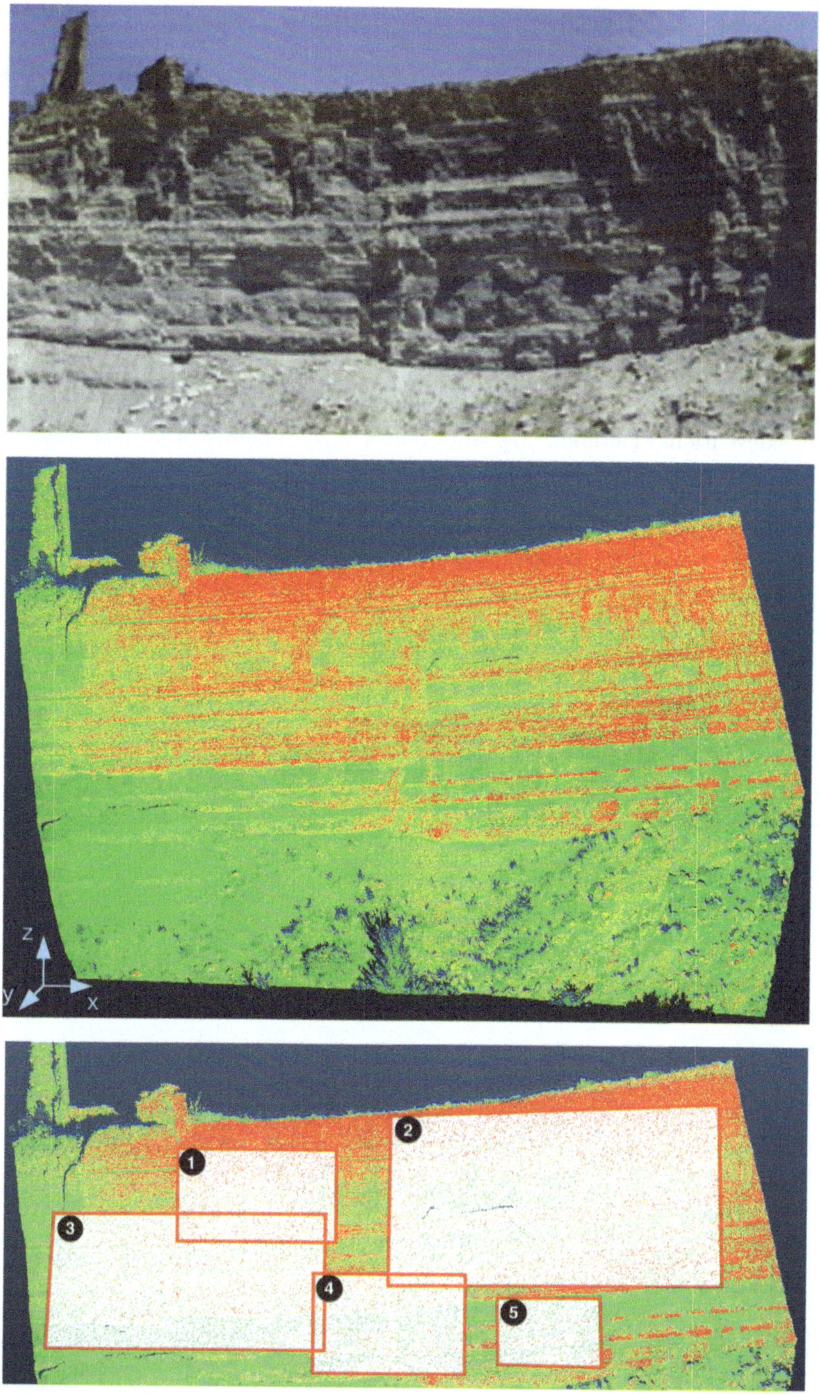

Fig. 3. 3D point cloud dataset (middle) scanned from the Puigcercos cliff (top), Catalonia, Spain [1,3,12], and five regions used in our simulations (bottom).

4.3 Experimental Results

Comparisons of data points detected in rockfall events between the proposed method and the conventional method [12] are shown in Table 1. It can be seen that, in a total of 34 events, the proposed method could identify 30 events, and the conventional method could detect 27 events. The conventional method is slightly better than the proposed method in this aspect. The proposed method could not detect some events, except the event c in region 3, because the sizes of the simulated hemispheres are smaller than the box's size w. In other words, the smallest event the proposed method can detect has a size of w. These results are not beyond the expectation and can be solved by reducing the box's size.

Table 1. Comparison of data points detected in rockfall events between the proposed and the conventional methods [12]. Note that nFP is a number of false-positive points. Note also that the numbers in parentheses are radii.

Region 1

	a(25)	b(20)	c(7)	d(20)	e(60)	f*(50)	g(25)	nFP
Ground truth	153	97	14	111	768	1060	160	0
Conventional	161	100	13	129	774	1084	174	15
Proposed	97	65	0	95	501	948	121	7

Region 2

	a*(60)	b(40)	c(15)	d*(40)	e(25)	f(40)	g(20)	h(30)	i(60)	nFP
Ground truth	1087	548	61	471	198	568	144	308	1546	0
Conventional	1099	554	60	476	203	576	145	313	1568	16
Proposed	173	245	11	0	77	284	94	155	1389	0

Region 3

	a(7)	b(18)	c(10)	d(20)	e*(50)	f(25)	g(40)	nFP
Ground truth	8	81	27	100	769	163	482	0
Conventional	0	80	25	102	776	170	487	56
Proposed	0	62	0	42	420	85	294	0

Region 4

	a(3)	b(25)	c(40)	d(40)	e*(40)	f(25)	g(20)	nFP
Ground truth	3	197	501	520	859	194	135	0
Conventional	0	204	510	530	877	199	137	4
Proposed	0	165	462	497	775	161	126	0

Region 5

	a(4)	b(5)	c(14)	d*(25)	nFP
Ground truth	7	10	74	234	0
Conventional	0	0	76	243	0
Proposed	0	0	48	202	0

However, in terms of false-positive detection, the proposed method is better, i.e., the proposed method infrequently identified rockfall when there is no rockfall compared to the conventional method. Example of detected rockfall events is shown in Fig. 2.

Figure 4 shows the comparison of the computational time complexity between the proposed method and the conventional method when the number of data points increases. It can be seen that the proposed method is faster.

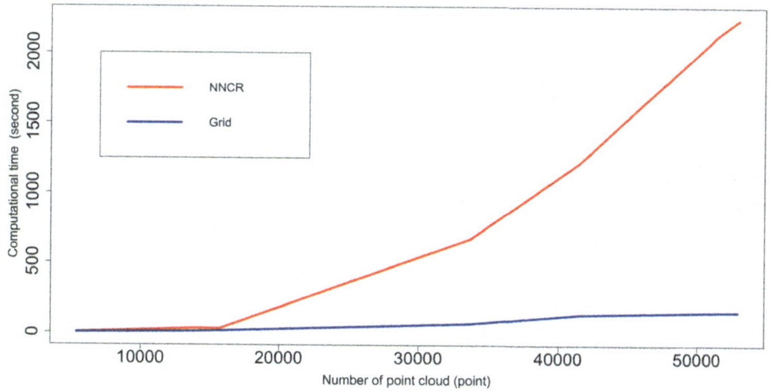

Fig. 4. Comparison of the computational time between the proposed method and the conventional method [12].

5 Discussion

In this section, we discuss the potential of using grid density to estimate the parameters of DBSCAN. As mentioned in Sect. 2.2, the DBSCAN algorithm requires two reasonable parameters to cluster data points effectively. Those parameters are ε and $\mathtt{minPts}$. The parameter ε is a distance to a point p that defines a set $N_\varepsilon(p)$, and the parameter $\mathtt{minPts}$ is an integer that defines the core point, the point that lies in high-density areas. Hence, if the smallest distance between two points on a laser line can be approximated (e.g., by the instrument's resolution), and the size w of the box of the grid is set to the laser line scanner resolution, the maximum number of points in the box (i.e., $c_{\mathtt{max}}$) can be reasonably used to approximate $\mathtt{minPts}$. That is, $\mathtt{minPts} = c_{\mathtt{max}}$, and $\varepsilon = w$. Figure 2 (f) shows an example of the result of applying DBSCAN with $c_{\mathtt{max}}$ and w to a surface-difference image. It can be seen that the resulting image is clearly better than the image in Fig. 2 (d). This issue will be investigated further. Note that this result was obtained under the condition that $[a, b]$ is $[1, 2]$.

Another critical issue worth discussing in this section is a relationship between the threshold used in the subtraction process by CloudCompare and the shifting interval $[a, b]$. It is clear that the threshold affects the number of points in the surface-difference image directly. In our simulation, we fixed $[a, b]$ to $[1, 2]$ to correspond to a possible situation for the laser line resolution of approximately 2 cm. We, therefore, fixed the threshold value to 1.8 cm. This threshold was set to this value because it resulted in a good performance for the conventional method [12], to which our proposed method is compared. The relationship between the threshold and the shifting interval should be studied further as well.

6 Conclusion

This paper proposed a method for rockfall detection based on DBSCAN with grid-density-based clutter removal from two data point clouds. The method consists of four steps: registration, subtraction, clutter removal, and DBSCAN. The paper's contribution is in providing a simple method for clutter removal in comparison with the conventional NNCR. The experimental results showed that the proposed grid-density-based clutter removal worked comparably to the NNCR, while the computational time complexity reduced reasonably. In addition, this paper suggested a straightforward means to estimate two parameters of the DBSCAN algorithm based on the laser line resolution and grid density.

Acknowledgement. This research is supported by Thailand Advanced Institute of Science and Technology (TAIST), National Science and Technology Development Agency (NSTDA), and Tokyo Institute of Technology under the TAIST-Tokyo Tech program. Also, this work has been partially supported by the ASEAN Committee on Science, Technology and Innovation (COSTI) under the ASEAN Plan of Action on Science, Technology and Innovation (APASTI) funding scheme and by e-Asia JRP funding scheme.

References

1. 3D Landslide Website. http://3d-landslide.com/
2. CloudCompare Website. https://www.danielgm.net/cc/
3. Abellán, A., Calvet, J., Vilaplana, J.M., Blanchard, J.: Detection and spatial prediction of rockfalls by means of terrestrial laser scanner monitoring. Geomorphology **119**(3–4), 162–171 (2010)
4. Bates, R.L., Jackson, J.A.: Glossary of geology (1987)
5. Bitelli, G., Dubbini, M., Zanutta, A.: Terrestrial laser scanning and digital photogrammetry techniques to monitor landslide bodies. Int. Arch. Photogramm. Remote Sens. Spat. Inf. Sci. **35**(B5), 246–251 (2004)
6. Byers, S., Raftery, A.E.: Nearest-neighbor clutter removal for estimating features in spatial point processes. J. Am. Stat. Assoc. **93**(442), 577–584 (1998)
7. Ester, M., Kriegel, H.P., Sander, J., Xu, X., et al.: A density-based algorithm for discovering clusters in large spatial databases with noise. In: Kdd, vol. 96, pp. 226–231 (1996)
8. Geertsema, M., Highland, L.: Landslides: human health effects (2011)
9. Jaboyedoff, M., et al.: Use of lidar in landslide investigations: a review. Nat. Hazards **61**(1), 5–28 (2012). https://doi.org/10.1007/s11069-010-9634-2

10. Micheletti, N., Tonini, M., Lane, S.N.: Geomorphological activity at a rock glacier front detected with a 3D density-based clustering algorithm. Geomorphology **278**, 287–297 (2017)
11. Schubert, E., Sander, J., Ester, M., Kriegel, H.P., Xu, X.: DBSCAN revisited, revisited: why and how you should (still) use DBSCAN. ACM Trans. Database Syst. (TODS) **42**(3), 1–21 (2017)
12. Tonini, M., Abellan, A.: Rockfall detection from terrestrial lidar point clouds: a clustering approach using R. J. Spat. Inf. Sci. **2014**(8), 95–110 (2014)

Humidity Sensor Drift Detection and Correction Based on a Kalman Filter with an Artificial Neural Network for Commercial Cultivation of Tropical Orchids

Kraithep Sirisanwannakul[1,2](✉), Nutchanon Siripool[1,2], Waree Kongprawechnon[2], Prachumpong Dangsakul[1], Udom Lewlomphaisarl[1], Seksun Sartsatit[1], Thanika Duangtanoo[1], Suthum Keerativittayanun[1], Wida Susanty Haji Suhaili[3], Yasunori Owada[4], Khin Than Mya[5], Sharifah H. S. Ariffin[6], and Jessada Karnjana[1]

[1] NECTEC, National Science and Technology Development Agency, 112 Thailand Science Park, Khlong Luang, Pathum Thani 12120, Thailand
prachumpong.dan@nectec.or.th
[2] Sirindhorn International Institute of Technology, Thammasat University, 131 Moo 5, Tiwanon Rd., Bangkadi, Muang, Pathum Thani 12000, Thailand
6022792111@g.siit.tu.ac.th
[3] Universiti Teknologi Brunei, Jalan Tungku Link, BE1410 Gadong, Brunei Darussalam
[4] National Institute of Information and Communications Technology, 2-1-3 Katahira, Aoba-ku, Sendai, Miyagi 980-0812, Japan
[5] University of Computer Studies, Yangon, No (4) Main Road, Shawe Pyi Thar Township, Yangon, Myanmar
khinthanmya@ucsy.edu.mm
[6] Universiti Teknologi Malaysia, Johor Bahru 81310, Johor Bahru, Johor, Malaysia

Abstract. Polymer dielectric-based humidity sensors used in the orchid greenhouse monitoring system usually work improperly after continuously being used in a high humid condition for some time (e.g., after eight months). This problem, called sensor drift, has been broadly observed. This paper proposes a simple data-driven technique based on a Kalman filter with an artificial neural network to detect the drift and correct data. The combination of two proposed measures based on the L^1 distance and the cosine similarity is used to determine the sensor's status, which is later used to adjust the Kalman gain accordingly. That is, when the sensor malfunctions, the gain is biased toward the prediction. When the sensor is in the normal status, the gain is biased toward the measurement. When the sensor drift is detected, the gain varies in between the prediction and the measurement. The experimental results show that the proposed method could reduce the accumulated mean absolute deviation by approximately 55.66%.

W. S. H. Suhaili et al. (Eds.): CIIS 2020, AISC 1321, pp. 140–149, 2021.
https://doi.org/10.1007/978-3-030-68133-3_14

Keywords: Kalman filtering · Neural network · Orchid greenhouse · Sensor drift detection and correction

1 Introduction

Smart agriculture has increasingly become an essential part of improving the quality and quantity of agricultural products [7]. The intelligent farm system basically aims to monitor and control the environmental factors necessary for plant growth to match the plant's needs [2,11]. For tropical orchid farming, where Thailand is one of the world's biggest contributors [1], the critical environmental factors consist of temperature, humidity, nutrients, air quality, and light intensity [10]. One of the demanding factors in the automatic monitoring system is air humidity because the humidity sensor tends to degrade when used for some time and continuously exposed under a high humidity condition [3]. This sensor's behavior is called the sensor drift [4]. Thus, to avoid problems due to the sensor drift, the sensor deployed in the system needs to be checked regularly and to be replaced if it malfunctions. Therefore, the smart agricultural system should detect such drift and correct data read from the drifting sensor automatically before it is replaced by the farmer.

Recently, some techniques have been proposed for sensor drift detection and correction [5,6,8]. For example, Dheeraj *et al.* proposed an automatic sensor drift detection and correction method based on spatial kriging and Kalman filtering [5]. They showed that the proposed method is robust against drifting and outperforms the traditional averaging-based interpolation methods [5]. However, one limitation of the kriging interpolation is that the accuracy drops when the number of data points is small. Zhan *et al.* applied a clustering method and a Kalman filter to solve the same problem [6]. Their approach does not require a densely deployed assumption and does not need prior knowledge of data models [6]. Their results showed that the proposed technique can improve the trustworthiness of collected data and is robust in a high-noise scenario [6]. However, the computational time complexity is expectedly enormous due to the singular value decomposition (SVD) used in their framework. Punit *et al.* proposed a method based on the Bayesian maximum entropy and an interacting-multiple model for drift detection and correction [8]. Their proposed method showed superiority to the Kalman-based methods [8]. It should be noted that, in their approach, they assumed that some sensors are high-precision sensors and do not drift. Thus, the problem formulation is different from the previous two mentioned methods to some degree. However, all of these previous methods assume a large-scale network. They make use of data from neighbors to detect the drift and correct the data. The large-scale-network approaches might not be appropriate well with the system deployed for an orchid greenhouse. Therefore, in this paper, we propose a new drift detection and correction method based on a Kalman filter that works with an artificial neural network without using neighbors' data.

The rest of this paper is organized as follows. Section 2 describes an overview of a monitoring system deployed in an experiment orchid greenhouse and the one-dimensional Kalman filtering algorithm. Section 3 provides detail of the proposed method. Experimental results and performance evaluation are reported in Sect. 4. The discussion regarding our findings is made in Sect. 5, and Sect. 6 summarizes this work.

2 Background

Two pieces of background information are provided in this section: the overview of the monitoring system used in this work and the Kalman filtering algorithm.

2.1 Overview of an Orchid Greenhouse Monitoring System

The greenhouse monitoring system used to collect data used in this work was installed at a commercial greenhouse in Ratchaburi, Thailand. The size of the greenhouse is 6 by 20 m^2. The monitoring system consists of many humidity sensors, temperature sensors, data loggers, a gateway, a weather station, and actuator controllers. Each sensor is located and covers an area of approximately 2 by 2.5 m^2. The data loggers collect data from the sensors and send them to the gateway through a wireless communication protocol, namely XBee. Then, the data are sent from the gateway to a database server via the Internet. The humidity sensor used in the system, namely SHT31, is of the capacitive type. We found that some sensors work improperly after eight months. For example, data are noisier and drifted from the actual values. This is in part due to the environment under which they are exposed. For instance, spraying liquid such as water (for controlling humidity) or insecticide chemicals constantly for some time could cause failure in relative humidity measurement. In such a high humid condition, this problem with the polymer dielectric-based sensor has been broadly observed [3].

2.2 One-Dimensional Kalman Filtering

There are two stages in performing the Kalman filtering algorithm: prediction and measurement update [9]. The core assumption of the algorithm is that a linear system can be described by one state equation and one measurement equation, as formulated by the following equations.

$$x_k = Ax_{k-1} + Bu_{k-1} + w_{k-1}, \tag{1}$$

and

$$y_k = Cx_k + v_k, \tag{2}$$

where x_k is the state at time k, u_{k-1} is an input at time $k-1$, w_{k-1} is a zero-mean process noise with a variance of Q at time $k-1$, y_k is a measured output at time k, v_k is a measurement white noise with a variance of R at time k, and A, B, and C are three factors concerning the state transition, the control input, and the relationship between the measurement and the state, respectively.

The prediction stage consists of two steps as follows. First, the algorithm predicts *a priori* state estimate $\hat{x}_k^-$ from the previous state estimate $\hat{x}_{k-1}$ by using the equation

$$\hat{x}_k^- = A\hat{x}_{k-1} + Bu_{k-1}. \tag{3}$$

Note that, for $k=1$, $\hat{x}_0$ is the initial state.

Second, the algorithm calculates *a priori* error variance from the equation

$$P_k^- = AP_{k-1}A + Q, \tag{4}$$

where P_{k-1} is *a posteriori* error variance. Note that, for $k=1$, P_0 is the initial error variance.

The measurement update stage consists of three steps as follows. First, the algorithm computes the Kalman gain K from

$$K = \frac{P_k^- C}{CP_k^- C + R}. \tag{5}$$

Second, the algorithm estimates the predicted state estimate $\hat{x}_k$ by using the following equation

$$\hat{x}_k = \hat{x}_k^- + K(y_k - C\hat{x}_k^-). \tag{6}$$

Third, the algorithm updates the state error variance by using the equation

$$P_k = (1 - KC)P_k^-. \tag{7}$$

3 Proposed Method

The goal of this work is to detect the sensor drift and estimate the actual humidity value based on a Kalman filter with a neural network, as shown in Fig. 1(a). The Kalman filter requires a model for the prediction stage and data for the measurement update stage. In our proposed method, the data read by the sensor at the location of interest are used as the measurement in the measurement update stage, and the neural network is used as the model for the prediction stage. The proposed method consists of two parts: detection and correction. Both parts use the same artificial neural network to their advantage.

The artificial neural network used in this work is of the feedforward class, called multi-layer perception. It aims to estimate the humidity value varying with time in a controlled environment. Thus, the network has one input (time) and one output (humidity value). A shallow neural network with two hidden layers is implemented. The numbers of nodes of the hidden layers are 20 and 20, respectively. The structure of the network is shown in Fig. 1(b). The hyperbolic tangent sigmoid transfer function is deployed as the activation function. The Levenberg-Marquardt algorithm is used to solve the non-linear least squares problem. The gradient descent with momentum weight with bias learning function is used as the adaptive learning function. The mean squared error is used to measure the performance of the network.

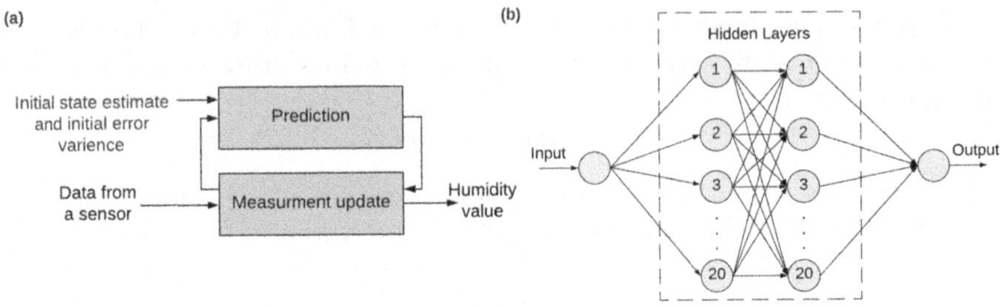

Fig. 1. Proposed method: (a) the diagram of Kalman filter with an artificial neural network and (b) the structure of neural network.

For the detection part, two measures Θ and Ω are defined based on the cosine similarity and the L^1 distance to detect the drift of a sensor. Let X_i be a vector of humidity values $x_{i,j}$ read from a sensor in a day i at time j. Hence, for a day that data are read every five minutes, $X_i = [x_{i,1}\ x_{i,2}\ ...\ x_{i,288}]^{\mathrm{T}}$. Let $\bar{X}$ be a vector of values $\bar{x}_j$ from the feedforward neural network at time j, i.e., $\bar{X} = [\bar{x}_1\ \bar{x}_2\ ...\ \bar{x}_{288}]^{\mathrm{T}}$.

The measure Θ is defined by

$$\Theta = -\log_{10}(1 - \cos\theta), \tag{8}$$

where the θ is the angle between X_i and $\bar{X}$, and $\cos\theta$ is calculated from the following equation.

$$\cos\theta = \frac{X_i \cdot \bar{X}}{\|X_i\|\|\bar{X}\|} = \frac{\sum_{j=1}^{288} x_{i,j}\,\bar{x}_j}{\sqrt{\sum_{j=1}^{288} x_{i,j}^2}\sqrt{\sum_{j=1}^{288} \bar{x}_j^2}}. \tag{9}$$

The measure Ω is defined by

$$\Omega = \log_{10}(d_1(X_i, \bar{X})), \tag{10}$$

where $d_1(X_i, \bar{X})$ is the L^1 distance between X_i and $\bar{X}$, which is computed from

$$d_1(X_i, \bar{X}) = \sum_{j=1}^{288} |x_{i,j} - \bar{x}_j|. \tag{11}$$

From our observation, as shown in Fig. 2, we found that X_i can be classified into three groups (i.e., normal, drift, and malfunction) based on sensor's status. Also, Θ and Ω can be used to determine the status. For the normal group, $\Omega \leq 3$ and $\Theta \geq 2.5$. In this case, the Kalman gain is biased toward the measurement by setting that $Q = 1$. For the drift group, $\Omega > 3$ and $\Theta \geq 2.5$. In this case, the

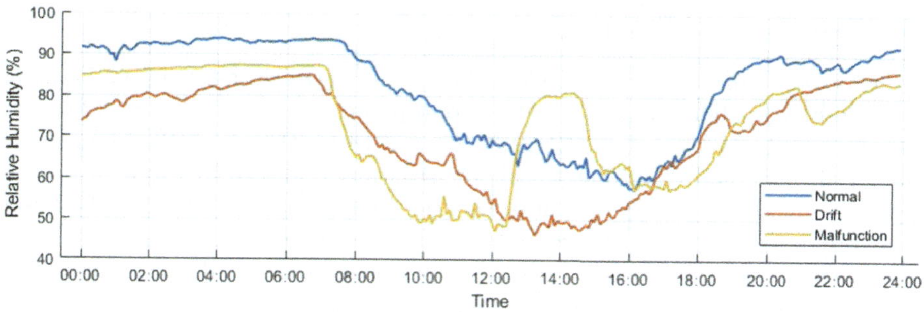

Fig. 2. Example of three sensor groups.

Kalman gain weights between the measurement and prediction by calculating the percentage of believing (β) in prediction from the equation

$$\beta = 5.88 + 58.82 \left(\frac{\Omega}{\Theta} \right). \tag{12}$$

The value of β is used to set the value of Q in such a way that the β values varying from 50% to 100% are linearly mapped to the Q values from 0.9 to 0.7, respectively. For the malfunction group, $\Theta < 2.5$, and the Kalman gain is biased toward the prediction model by setting that $Q = 0$.

For the correction part, the proposed method formulates the state equation of the Kalman filter as follows.

$$x_k = x_{k-1} + w_{k-1}, \tag{13}$$

where x_k is the estimated humidity value, which is the corrected value, at time k and w_{k-1} is the process noise at time $k-1$. The measurement equation is

$$y_k = x_k + v_k, \tag{14}$$

where y_k is the data read from the sensor of interest at time k and v_k is the measurement noise at time k. The predicted (*a priori*) state estimate $\hat{x}_k^-$ is then computed by

$$\hat{x}_k^- = \hat{x}_{k-1}, \tag{15}$$

where $\hat{x}_{k-1}$ is the previous state estimate that is determined by the artificial neural network, i.e., $\hat{x}_{k-1}$ is an element of $\bar{X}$.

4 Experiments and Results

We conducted experiments to verify that a Kalman filter with a feedforward neural network can be applied to detect and correct sensor drift. The details of our experiments, their setup, and results are provided in the following subsections.

4.1 Experimental Setup

In this work, humidity data are collected and sent to the database every 5 min, as shown in Fig. 3. Therefore, there are 288 datapoints a day. The duration for data analysis is 314 days, i.e., from 21 September 2018 to 31 July 2019. The general pattern of data read from one sensor for 314 days is shown in Fig. 4. It can be seen that the drift has occurred after the 279th day, as marked by a dashed line. Therefore, in our experiment, we applied the proposed method to detect the drift and to correct sensor's values from the 280th day to the 314th day. Also, data from the first 60 days were use to train the neural network since we assumed that the sensor worked properly during those first 60 days. The data predicted ($\bar{X}$) by the network are shown in Fig. 3, and the average root-mean-square error (RMSE) between $\bar{X}$ and X_j for $j = 61$ to 279 is 6.74 (%RH).

To evaluate the proposed method, the accumulated mean absolute deviation (accumulated MAD) from the ith day to the kth day ($\Delta_{i,k}$) is used, and it is defined by the following equation.

$$\Delta_{i,k} = \frac{1}{288}\sum_{l=i}^{k}\sum_{j=1}^{288} |x_{i,j} - \bar{x}_j| . \tag{16}$$

4.2 Experimental Result

The comparison of the accumulated drifts $\Delta_{i,k}$, for $i = 280$ and $k = 280$ to 314, is shown in Fig. 5. It can be seen that, at the 314th day, the proposed method could reduce the accumulated drift by approximately 55.66%.

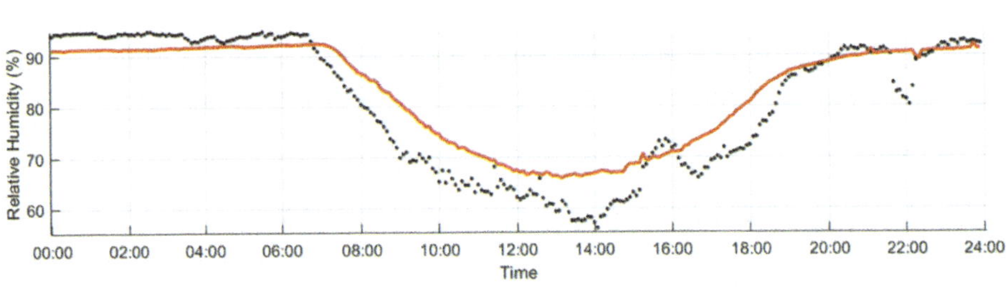

Fig. 3. Example of data in one day (dotted line) and predicted data (solid line).

Figure 5 shows the accumulated MAD values $\Delta_{i,k}$ from the proposed method, compared to those from data read from the sensor in the month the sensor malfunctioned. The MAD values during that period in two cases (with and without the proposed method) are 3.27 and 7.51, respectively. Therefore, the proposed Kalman filter with a neural network could considerably reduce the drift of the humidity sensor.

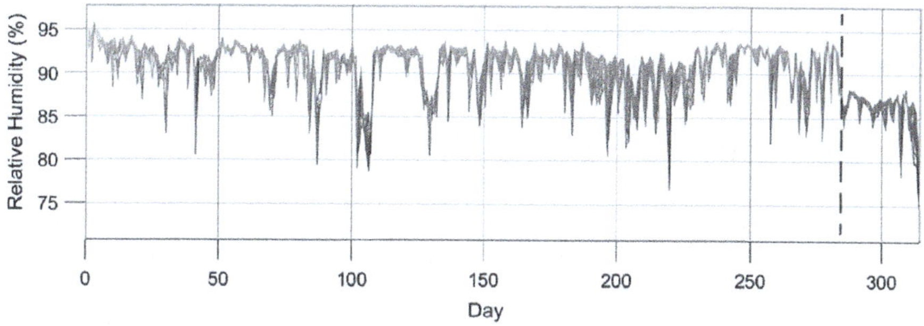

Fig. 4. Pattern of data read from one sensor for 314 days.

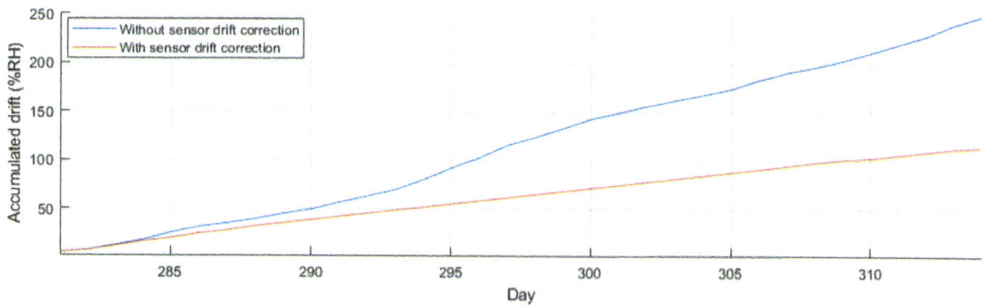

Fig. 5. Comparison of accumulated drifts $\Delta_{i,k}$ in two cases: with the proposed sensor drift detection and correction and without the proposed method.

5 Discussion

The experiments described so far has a gap due to the lack of ground-truth data. Thus, we discuss how large the difference between values from the prediction based on the neural network and those of the ground truth is. Also, this section discusses how to reduce that difference further.

We conducted two preliminary experiments to estimate data at one location from 21 September 2018 to 31 December 2018. In the one experiment, a Kalman filter with kriging interpolation was used as the estimator. In another experiment, a Kalman filter with bicubic interpolation was used. When the values read from the sensor were compared with estimated ones, we found that the average RMSEs were 1.32 (%RH) and 1.33 (%RH), respectively. However, the average RMSE value between predicted data from the neural network and those from the sensor during the same period was 5.41 (%RH), which is larger than the Kalman with kriging and the Kalman with bicubic interpolation. Both kriging and bicubic interpolations make use of nearby sensors' data to their advantage. Therefore, we expect that, if the Kalman filter's prediction stage can access and use neighbors' data (e.g., the neural network is trained not only by data at the positions of interest but also by data from surrounding sensors), the average RMSE should reduce. As a consequence, the overall performance will be improved. Last but not least, to fill the gap concerning the lack of ground truth, the further experiment with some reliable and trustable sensor will be conducted in the future.

6 Conclusion

This paper proposed a method for sensor drift detection and correction in humidity measurement for tropical orchids commercial cultivation. The method is based on a Kalman filtering algorithm, of which its prediction is constructed from a simple artificial neural network. Two measures based on the L^1 distance and the cosine similarity were proposed to detect the sensor drift and determine the sensor's status, which is then used to adjust the Kalman gain correspondingly. That is, when the sensor malfunctions, the Kalman gain is biased toward the prediction; otherwise, it is biased toward the measurement. The experimental results showed that the proposed method could reduce the accumulated mean absolute deviation by approximately 55.66%.

Acknowledgement. This work is the output of an ASEAN IVO (http://www.nict.go.jp/en/asean_ivo/index.html.) project, titled 'A Mesh-topological, Low-power Wireless Network Platform for a Smart Watering System,' and partially financially supported by NICT (http://www.nict.go.jp/en/index.html.). The authors of this paper would like to express their sincere gratitude to Thai Orchids Co., Ltd., for the experiment greenhouse.

References

1. Department of International Trade Promotion.: The orchid world situation (2018). https://www.ditp.go.th/contents_attach/539560/539560.pdf. Accessed 27 June 2020
2. Du, X., Wang, J., Ji, P., Gan, K.: Design and implement of wireless measure and control system for greenhouse. In: Proceedings of the 30th Chinese Control Conference, pp. 4572–4575. IEEE (2011)
3. Griesel, S., Theel, M., Niemand, H., Lanzinger, E.: Acceptance test procedure for capacitive humidity sensors in saturated conditions. WMO CIMO TECO-2012, Brussels, Belgium, pp. 1–7 (2012)
4. Holmberg, M., Artursson, T.: Drift compensation, standards, and calibration methods. In: Handbook of Machine Olfaction: Electronic Nose Technology, pp. 325–346 (2002)
5. Kumar, D., Rajasegarar, S., Palaniswami, M.: Automatic sensor drift detection and correction using spatial Kriging and Kalman filtering. In: 2013 IEEE International Conference on Distributed Computing in Sensor Systems, pp. 183–190. IEEE (2013)
6. Li, Z., Wang, Y., Yang, A., Yang, H.: Drift detection and calibration of sensor networks. In: 2015 International Conference on Wireless Communications and Signal Processing (WCSP), pp. 1–6. IEEE (2015)
7. McBratney, A., Whelan, B., Ancev, T., Bouma, J.: Future directions of precision agriculture. Precision Agric. **6**(1), 7–23 (2005)

8. Rathore, P., Kumar, D., Rajasegarar, S., Palaniswami, M.: Bayesian maximum entropy and interacting multiple model based automatic sensor drift detection and correction in an IoT environment. In: 2018 IEEE 4th World Forum on Internet of Things (WF-IoT), pp. 598–603. IEEE (2018)
9. Rhudy, M.B., Salguero, R.A., Holappa, K.: A Kalman filtering tutorial for undergraduate students. Int. J. Comput. Sci. Eng. Surv. **8**(1), 1–9 (2017)
10. Stuckey, I.H.: Environmental factors and the growth of native orchids. Am. J. Bot. **54**(2), 232–241 (1967)
11. Xing, X., Song, J., Lin, L., Tian, M., Lei, Z.: Development of intelligent information monitoring system in greenhouse based on wireless sensor network. In: 2017 4th International Conference on Information Science and Control Engineering (ICISCE), pp. 970–974. IEEE (2017)

Improved Parking Space Recognition via Grassmannian Deep Stacking Network with Illumination Correction

Tee Connie[1](✉), Michael Kah Ong Goh[1], Voon Chet Koo[1], Ken T. Murata[2], and Somnuk Phon-Amnuaisuk[3]

[1] Multimedia University, Jalan Ayer Keroh Lama, 75450 Melaka, Malaysia
{tee.connie,michael.goh,vckoo}@mmu.edu.my

[2] National Institute of Information and Communications Technology, Tokyo, Japan
ken.murata@nict.go.jp

[3] Universiti Teknologi Brunei, Jalan Tungku Link Gadong, Bandar Seri Begawan 1410, Brunei Darussalam
somnuk.phonamnuaisuk@utb.edu.bn

Abstract. It has become increasingly difficult to quickly locate a free parking space with the growing number of private vehicles. Many parking space management solutions have been proposed. Vision-based methods are among the approaches that have received great attention due to the widespread use of surveillance cameras in parking areas. Although promising results have been reported for vision-based methods, these methods generally suffer when good quality images are not available. The performance of vision-based methods drops under conditions like low illumination. In this paper, an approach coined as Grassmannian Deep Stacking Network with Illumination Correction (GDSN-IC) is presented. The proposed method enhances the illumination map of an image before feeding it to a Grassmannian Deep Stacking Network for parking space availability prediction. Experiments on two public datasets validate the effectiveness of the proposed approach.

Keywords: Parking space recognition · Grassmann manifold · Deep Stacking Network · Illumination Correction

1 Introduction

Due to the progressively rising number of private vehicles in urban cities, finding a vacant parking space has become a major problem. Studies have revealed that around 30% of traffic congestion in urban cities are caused by vehicles searching for parking spaces [1]. Cruising for parking space is not only energy and time-consuming, but also causes environment pollution due to vehicle emissions. If the search for a vacant parking space turns out in vain, a frustrated driver may simply park his/her vehicle in an unauthorized area which may cause accidents.

W. S. H. Suhaili et al. (Eds.): CIIS 2020, AISC 1321, pp. 150–159, 2021.
https://doi.org/10.1007/978-3-030-68133-3_15

Various parking space management systems have been proposed to address the vehicle parking problem. Some systems deploy a counter-based approach to count the number of vehicles entering and existing a parking area. Devices like gate-arm counters and inductive loop detectors are installed at the entrances and exits of the parking area to keep track of parking space occupancy. These devices are low-cost but such systems cannot tell the actual location of the vacant parking space. There are also sensor-based systems that install sensors like magnetic field sensor and ultrasonic sensor at each parking bay to detect the existence of a vehicle at the parking bay. Such sensor-based systems can tell the location of the vacant parking space but the installation cost is high. It is not feasible to install such sensor-based systems in parking areas like outdoor parking that involves a large number of parking spaces.

With the proliferation of surveillance cameras in public areas, it is not difficult to obtain video footages of the car park areas. This leads to an increase attention in vision-based car park management systems. However, vision-based systems are faced with a number of challenges among which illumination condition poses a major difficulty. This problem is particularly evident in outdoor parking where the same scene appears different under different lighting variations and weather conditions. It is difficult for a parking space detection method to work effectively under low-light density, for example night scene.

This paper reports our work in progress from the ASEAN-IVO project titled Event Analysis: Applications of computer vision and AI in the smart tourism industry. Parking space analysis is one of the objectives. In this paper, a method called Grassmannian Deep Stacking Network with Illumination Correction (GDSN-IC) is proposed to address the illumination problem in parking space recognition. A lighting assessment metric is first applied to estimate the illumination map of the image. After that, an illumination correction technique is used to enhance the image quality. A set of corrected car park images are modelled as a manifold-valued point and fed as input to a Grassmannian Deep Stacking Network for classification. Experiments conducted on two publicly available carpark datasets, namely PKLot [2] and CNRPark-Ext [3] datasets, validate the effectiveness of the proposed GDSN-IC method.

2 Literature Review

The existing parking space detection methods can be broadly categorized into: (1) Sensor-based approach, and (2) Vision-based approach. For sensor-based approach, various sensors like magnetic field sensor [4, 5], ultrasonic sensor [6], infra-red sensor [7] and light dependent resistor [8] and are used. As magnetic sensors are susceptible to interference from vehicles parked at adjacent parking spaces, Zhang et al. [4] proposed to fuse signals of adjacent sensors. The peak values of the signals were used to determine whether there was a vehicle at the parking space. Another parking occupancy detection method using magnetic sensor was presented in [5]. Some feature extraction techniques were used to obtain meaningful features from the magnetic signal. Distance discriminate analysis was then applied to classify the result as vacant or occupied. The use of ultrasonic sensor was investigated in [6] to detect the existence of a vehicle at a parking space. The study found that multiple echo functions could detect edges of the vehicle.

The obstacle detection function of infra-red sensors was deployed in [7] to detect vacant parking space. The study checked whether the IR receiver path was blocked. If blocked, it indicated the presence of a vehicle at the parking space. [8] presented a mechanism for vacant parking space detection using different sensors like magnetometer, ultrasound sensor, infra-red sensor and light dependent resistor. The study found that magnetometer sensors are the most reliable for vacant parking space detection. A combination of multiple sensors was also utilized [9] for vacant parking space detection. Ultrasonic sensor, magnetic sensor and infra-red sensor were tested in the study. The authors found that the combination of magnetic sensor and ultrasonic sensor yielded the best result. A summary of sensor-based methods is provided in Table 1.

Table 1. Summary of sensor-based methods.

Authors	Sensor Type	Approach	Performance
Zhang et al. [4]	Magnetic field sensor	Adjacent sensors signal fusion	Accuracy = 99.54%
Zhang et al. [5]	Magnetic field sensor	Distance discriminate analysis	Accuracy = 98%
Park et al. [6]	Ultrasonic sensor	Multiple echo functions	Error measurement ~16 cm
Srikurinji et al. [7]	Infra-red sensor	Path blocking detector	Accuracy above 90%
Narayanan et al. [8]	Magnetometer, ultrasound, infra-red and light dependent resistor	Object detection	Accuracy = 100%
Jindaprakai et al. [9]	Ultrasonic, magnetic and infra-red sensor	Object detection	Accuracy = 100%

The vision-based approach, on the other hand, does not rely on sensors to detect the existence of a vehicle in a parking space. Instead, it analyses images/scenes of the parking area to determine an available parking space. The earlier vision-based methods used texture information [10], corner detection and color information [11] to detect vacant parking space. 3D modelling technique had also been proposed to model the parking lot scene with HoG features [12]. Background subtraction using GMM and transient map had also applied for vacant parking area estimation [13]. A similar method was presented in [14], added with Kalman filter tracking. Recently, bag of feature and background subtraction techniques were also introduced by [15] to detect delimited and non-delimited parking spaces.

Due to the success of deep learning technology, a number of deep learning architectures have been proposed to solve parking space detection problem. Convolution Neural Network (CNN) had been proposed in [3] for parking occupancy detection. The authors presented a smaller version of AlexNet called mAlexNet. Deep CNN and a multi-task contrastive network with spatial transform was presented in [16]. The proposed method could solve the inter-object occlusion in parking areas. A mask R-CNN model was built in [17] for locating free parking space. The proposed method also detected two wheelers to optimize parking space occupancy. Deep learning-based object detection method like YOLO was also deployed to recognize parking occupancy [18]. The proposed method was embedded in IoT devices and connected directly to the existing street surveillance.

Table 2. Summary of vision-based methods.

Authors	Method	Dataset	Performance
True [11]	Color histogram &Harris corner detection	Self-collected dataset	Accuracy = 94%
Huang et al. [12]	3D parking lot model	Self-collected dataset	Average accuracy = 99.5%
Postigo et al. [13]	GMM background subtraction	Self-collected dataset	Avg False negative = 9.32%
Marmol et al. [14]	Background subtraction & Kalman tracking	QuickSpotDB	Avg detection rate = 98.8%
Varghese et al. [15]	Bag of feature & background subtraction	PKLot and CNRPark	Avg accuracy = 96.53%
Amato et al. [3]	Convolutional Neural Network	PKLot and CNRPark-EXT	Accuracy > 90%
Vu et al. [16]	CNN and Multi-task Contrastive Network	Self-collected dataset	Accuracy = 99.42%
Sairam et al. [17]	Mask R-CNN	PKLot	Accuracy = 98.4%
Chen et al. [18]	YOLO and MobileNet	CNRPark + EXT	Accuracy = 99%

3 Proposed Method

The proposed GDSN-IC approach is made up of several components. Given an input image, an illumination map estimation module is first applied to estimate the image lighting composition. After that, an illumination correction module is called to improve the lighting effect. Next, the set of enhanced images are fed to a Grassmannian Deep Stacking Network (GDSN) to obtain a predicted output whether the parking space is vacant or occupied. The block diagram of the proposed approach is illustrated in Fig. 1.

3.1 Illumination Map Estimation and Illumination Correction

For correct recognition of available parking spaces, a clear and highly visible image is desirable. Therefore, it is important to apply illumination correction method on low-light

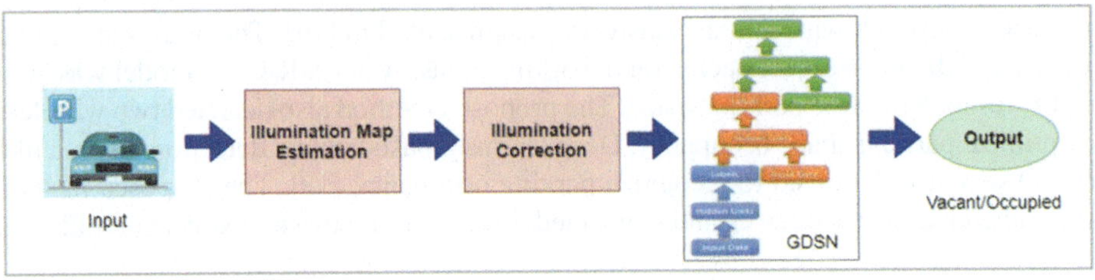

Fig. 1. Block diagram of the proposed approach.

images to bring out details of the parking scene. Given an input image **x**, the illumination map of the image **m** can first estimated using [19],

$$\mathbf{m}(i) \leftarrow \max_{C \in \{R,G,B\}} \mathbf{x}^C(i) \tag{1}$$

The recovery **r** is given by,

$$\mathbf{r}(i) = \frac{\mathbf{x}(i)}{\mathbf{max}_C\, \mathbf{x}^C(i) + \varepsilon} \tag{2}$$

where ε is a very small constant to avoid divide by zero. After that, the illumination map **m** can be corrected via Gamma correction $\widehat{\mathbf{m}} \leftarrow \mathbf{m}^{\gamma}$. The image **x** is then enhanced through $\widehat{\mathbf{m}}$ using,

$$\hat{\mathbf{x}} = \mathbf{r} \circ \widehat{\mathbf{m}} \tag{3}$$

where $\hat{\mathbf{x}}$ is the enhanced image and $\circ$ refers to element-wise multiplication.

3.2 Grassmannian Deep Stacking Network (GDSN)

In this paper, a deep learning approach formulated on the Grassmann manifold coined as Grassmannian Deep Stacking Network (GDSN) is proposed. Most existing deep learning systems are developed for Euclidean spaces considering linear structure (vector spaces). This work goes beyond these representations and extend the stacking-based deep neural network [20] to a non-linear space of curved Grassmannian manifold. Kernelised orthonormal random projection [21] is used to embed the manifold structure in a high-dimensional Reproducing Kernel Hilbert Space (RKHS). The overall framework of GDSN is shown in Fig. 2.

Grassmann Points Formulation. Given an image, **x**, represented as a vector containing d elements (the number of pixels), the image can be viewed as a point in the image space. In this study involving video-based data, there are a set of k images which can be grouped to form an d-by-k matrix $\mathbf{X} = \{\mathbf{x}_1, \mathbf{x}_2, \ldots, \mathbf{x}_k\}$. As an orthonormal basis can always be associated with the column space of the matrix, e.g. through QR-decomposition or Singular Value Decomposition [22], a k-dimensional subspace $\mathcal{V}$ can be formed for matrix **X**.

When the Grassmann manifold is realized as a submanifold of a projective space, points in the Grassmann manifold are equivalent to classes of d-by-k orthonormal matrices [22]. Therefore, $\mathcal{V}$ can be encoded naturally as a point on the Grassmann manifold. In this regard, performing classification of the image sets is equivalent to performing classification of points on the Grassmann manifold.

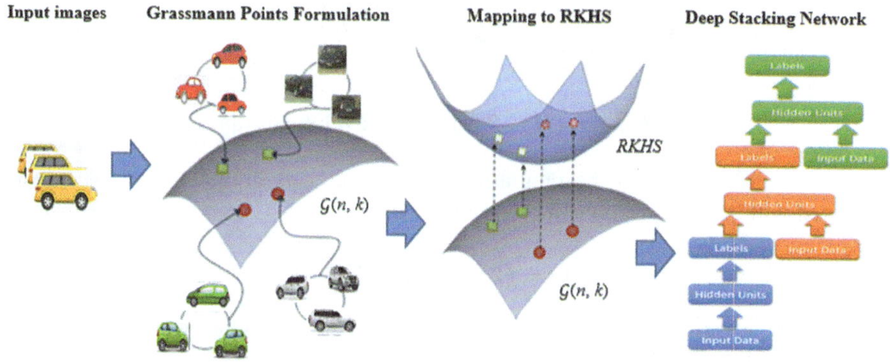

Fig. 2. Overall framework of GDSN.

Random Projection via Orthonormal Hyperplanes. To apply random projection on the Grassmann space, a subset of the training exemplars $\mathcal{S} = \{\mathcal{V}_1, \ldots, \mathcal{V}_p\}$are randomly selected from m training data, where $p \ll m$. After that, the kernelized points $\varnothing(\mathcal{V}_i)_{i=1}^{p}$ are arranged to form a matrix $\mathcal{O} = \left[\varnothing(\mathcal{V}_1), \ldots, \varnothing(\mathcal{V}_p)\right]$. To ensure the hyperplanes in the random projection space are independent, $\mathcal{O}$ needs to be orthogonalized. To achieve this, the standard QR decomposition [23] can be used:

$$\mathcal{O} = \left[\varnothing(\mathcal{V}_1), \ldots, \varnothing(\mathcal{V}_p)\right] = \mathbf{Q}\widehat{\mathbf{R}} \tag{4}$$

where $\mathbf{Q}$ and $\widehat{\mathbf{R}}$ denote the orthogonal matrix and upper triangular matrix.

Projecting a data point $\varnothing(\mathcal{V}_1)$ into the orthogonal basis $\mathbf{Q}$ is equivalent to projecting the kernel matrix into the upper triangular matrix $\widehat{\mathbf{R}}$, where $\varnothing(\mathcal{V})^T\mathbf{Q} = \left[\mathbf{K}(\mathcal{V}, \mathcal{V}_1), \ldots, \mathbf{K}(\mathcal{V}, \mathcal{V}_p)\right]\widehat{\mathbf{R}}^{-1}$. To compute $\widehat{\mathbf{R}}$, only the kernel Gram matrix $\mathbf{K}_{\mathcal{S}}$is needed where $\mathbf{K}_{\mathcal{S}} = \hat{\mathbf{R}}^T\mathbf{R}$. Hence, Cholesky Factorisation [24] can be used to find the upper triangular matrix using the kernel Gram matrix $\mathbf{K}_{\mathcal{S}}$. In this way, multiplying the kernelized Grassmann points with $\widehat{\mathbf{R}}^{-1}$ is equivalent to projecting the points into a random projection space spanned by $\mathcal{S}$.

Construction of GDSN. The proposed deep stacking structure relies on ridge regression due to its simple and tractable analytic solution. Each layer of the network is made up of independent module which takes the output from the previous layer as the input. In this way, each layer learns a unique feature representation and pass it to the subsequent layer to learn a more exquisite feature.

Assume $\left\{\widetilde{\mathcal{X}}_i\right\}_{i=1}^{N}$ are the N Grassmann points projected into the orthonormal random projection space with the corresponding class labels y_i, where $\widetilde{\mathcal{X}}_i \in \mathbb{R}^p$ and $y_i \in \{0, 1\}^C$; C is the number of training classes. The minimization problem for the penalized residual sum of squares β is given by,

$$\beta(w)_{l2} = \sum\nolimits_{i=1}^{N}\left(y_i - \widehat{\sum\nolimits_{j=1}^{p}}\widetilde{\mathcal{X}}_{i,j}w_j\right)^2 + \lambda\sum\nolimits_{j=1}^{p}w_j^2 \tag{5}$$

Let $\mathcal{Z} = [\tilde{\mathcal{X}}_1, \ldots, \tilde{\mathcal{X}}_N] \in \mathbb{R}^{N \times p}$, and $\mathcal{Y} = [y_1, \ldots, y_N] \in \mathbb{R}^{N \times C}$, Eq. (5) can be rewritten as:

$$\beta(\mathbf{w})_{l2} = (\mathcal{Y} - \mathcal{Z}\mathbf{w})^T(\mathcal{Y} - \mathcal{Z}\mathbf{w}) + \lambda\|\mathbf{w}\|_2^2 \quad (6)$$

where $\mathbf{w} \in \mathbb{R}^{p \times C}$ and λ is the regularization parameter. A unique solution is always associated with $\beta(\mathbf{w})_{l2}$ which can be estimated as:

$$\mathbf{w} = (\mathcal{Z}^T\mathcal{Z} + \lambda\mathbf{I})^{-1}\mathcal{Z}\mathcal{Y} \quad (7)$$

where $\mathbf{I}$ is the identity matrix. During the evaluation stage, the class label $\mathcal{Y}_t$for a query input $\mathcal{Z}_t$is predicted with,

$$\mathcal{Y}_t = \mathbf{w}^T\mathcal{Z}_t \quad (8)$$

When training multiple layers of network, the output from a layer, $\tilde{\mathcal{Y}}_t$, is concatenated with the original data $\tilde{\mathcal{X}}_i$ to form the input to the subsequent layer.

4 Experiments and Discussions

4.1 Datasets

Two public datasets are used to testify the robustness of the proposed method. They are PKLot [2] and CNRPark [3] datasets. The PKLot dataset contains a large number of parking lot images labelled as “Occupied” or “Empty”. The images were captured at a resolution of 1280 × 720 from three camera views. The images came from two parking lots. Three weather conditions were included in the dataset namely sunny, cloudy and raining.

The CNRPark dataset contains two variations. CNRPark is a smaller dataset containing about 12,000 labeled images while CNRPark-EXT is an extension of CNRPark that consists of 145,000 labeled images. The images for CNRPark were captured from two view angles while the images for CNRPark-EXT were recorded using 9 view angles across different lighting conditions and occlusion patterns. A summary of the different car park datasets is provided in Table 3.

Table 3. Summary of car park datasets.

Datasets	Free spaces	Busy spaces	Totals
PKLot	337,780	358,119	695,899
CNRPark	4181	8403	12,584
CNRPark-EXT	65,658	79,307	144,965

4.2 System Evaluation

The set of parking lot images was randomly subsampled into two partitions where a partition was used as the training set while the other for testing. k-fold cross validation was applied in the tests. To form a Grassmann point, ten parking lot images was selected at random. The images were converted to grayscale images and normalized to 64×64 pixels. The elements for $\mathcal{V}$ were randomly selected from the training set. The canonical correlation kernel was chosen for $\mathbf{K}$. The regularization parameter λ was empirically determined in the tests.

There are two major components to be evaluated in the proposed method: (1) the dimension of the random projection space, p, and (2) different number of layers for the network. The output contains only two classes, namely vacant and occupied. The correct recognition rate (CRR) that counts the occurrence of correct prediction is used as the accuracy measure. Figure 3(a) depicts the results for using different values for p, the dimension of the projected space. Generally, the best performance is obtained when the value of p is set to around 650, which is about 80% of the original feature dimension. Using less dimension would increase computational efficiency, but this results in sacrifice of accuracy.

The next experiment is performed to evaluate the performance of the proposed GDSN-IC method. The results of using different number of layers for the network is shown in Fig. 3(b). The value of p was chosen at 650 in the experiment. Overall, using only two layers is sufficient to yield good results. Marginal performances are achieved by adding more layers but this results in increase in computation. The features learnt using a few layers are already sufficient to yield satisfactory results.

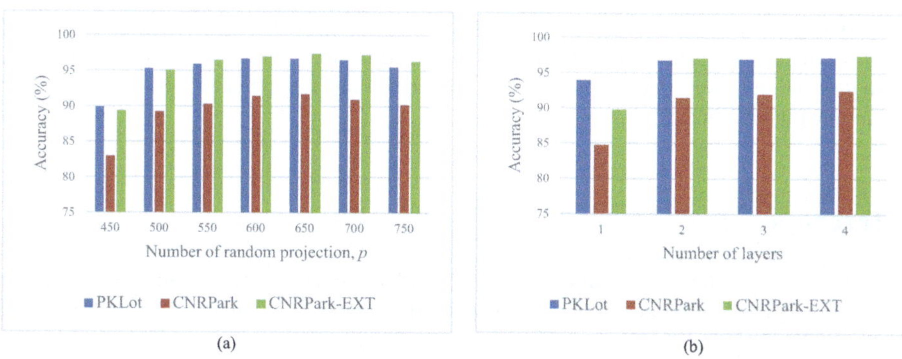

Fig. 3. Accuracy for varying number of (a) random projection, and (b) number of layers.

The performance of the proposed method is also compared with the other recent techniques. The performances of the different methods are shown in Table 4. The performance of GDSN-CI is comparable to the state-of-the-art methods. All the methods are based on deep learning principals. Nevertheless, GDSN-CI requires much lower computation time for training. There are at least two advantages of the proposed approach:

Table 4. Comparison with state-of-the-art techniques in terms of accuracy (%).

Datasets	PKLot	CNRPark	CNRPark-EXT
Varghese et al. [15]	96.59	91.24	96.59
Amato et al. [3]	95.88	90.71	95.88
GDSN-IC	96.80	91.50	97.10

(a) Allows a robust way to exploit the non-linear geometry of Grassmann manifold for deep analysis, and (b) Offers a scalable and analytical solution for building a deep architecture.

5 Conclusion

Search of parking space is a pressing problem in urban cities. This paper presents a method called GDSN-IC to address illumination problem in parking space recognition. The illumination map of the parking images is estimated and used to enhance the image quality. The proposed deep stacking network consists of a number of "shallow" modules that are combined to eventually form a deep architecture. Each shallow modules are trained independently which makes it possible to learn a convex solution for the deep networks. Experiments on the PKLots and CNRPark datasets validate the effectiveness of the proposed approach.

Acknowledgement. The ASEAN IVO (https://www.nict.go.jp/en/asean_ivo/index.html) project, Event Analysis: Applications of computer vision and AI in smart tourism industry, was involved in the production of the contents of this work and financially supported by NICT (https://www.nict.go.jp/en/index.html).

References

1. Tsiaras, C., Hofstetter, F., Liniger, S.: parkITsmart: minimization of Cruising for Parking. In: Presented at the 24th International Conference on Computer Communications and Networks (ICCCN) (2015)
2. Almeida, P.R.L., Oliveira, L.S., Britto, A.S., Silva, E.J., Koerich, A.L.: PKLot – a robust dataset for parking lot classification. Expert Syst. Appl. **42**, 4937–4949 (2015)
3. Amato, G., Carrara, F., Falchi, F., Gennaro, C., Meghini, C., Vairo, C.: Deep learning for decentralized parking lot occupancy detection. Expert Syst. Appl. **72**, 327–334 (2017)
4. Zhang, Z., Mao, X., Zhou, K., Yuan, H.: Collaborative sensing-based parking tracking system with wireless magnetic sensor network. IEEE Sens. J. **20**, 4859–4867 (2020)
5. Zhang, Z., Tao, M., Yuan, H.: A parking occupancy detection algorithm based on AMR Sensor. IEEE Sens. J. **15**, 1261–1269 (2015)
6. Park, W.-J., Kim, B.-S., Seo, D.-E., Kim, D.-S., Lee, K.-H.: Parking space detection using ultrasonic sensor in parking assistance system. In: 2008 IEEE Intelligent Vehicles Symposium, pp. 1039–1044 (2008)

7. Srikurinji, M.S., Prema, M.U., Sathya, M.S., Manivannan, M.P.: Smart Parking System Architecture using infrared detecter. paper/SMART-PARKING-SYSTEM-ARCHITECTURE-USING-INFRARED-Srikurinji-Prema/4bbca13aca486d5e392511767ea32c24e50ee84c. Accessed 26 Aug 2020
8. Narayanan, N.S., Shaji, F., Praseedom, J., Thiyagarajan, P.: Intelligent parking and power management system using sensors. In: 2019 International Conference on Vision Towards Emerging Trends in Communication and Networking (ViTECoN). pp. 1–7 (2019)
9. Jindaprakai, N., Nuchitprasitchai, S.: Intelligent parking system using multiple sensor detection. In: 2019 Research, Invention, and Innovation Congress (RI2C), pp. 1–4 (2019)
10. Lixia, W., Dalin, J.: A method of parking space detection based on image segmentation and LBP. In: 2012 Fourth International Conference on Multimedia Information Networking and Security, pp. 229–232 (2012)
11. True, N.: Vacant parking space detection in static images. Univ. of California San Diego. **17**, 659–662 (2007)
12. Huang, C.-C., Tai, Y.-S., Wang, S.-J.: Vacant parking space detection based on plane-based Bayesian hierarchical framework. IEEE Trans. Circuits Syst. Video Technol. **23**, 1598–1610 (2013)
13. del Postigo, C.G., Torres, J., Menéndez, J.M.: Vacant parking area estimation through background subtraction and transience map analysis. IET Intel. Transport Syst. **9**, 835–841 (2015)
14. Màrmol, E., Sevillano, X.: QuickSpot: a video analytics solution for on-street vacant parking spot detection. Multimed Tools Appl. **75**, 17711–17743 (2016)
15. Varghese, A., Sreelekha, G.: An efficient algorithm for detection of vacant spaces in delimited and non-delimited parking lots. IEEE Trans. Intell. Transp. Syst. 1–11 (2019)
16. Vu, H.T., Huang, C.-C.: Parking space status inference upon a deep CNN and multi-task contrastive network with spatial transform. IEEE Trans. Circuits Syst. Video Technol. **29**, 1194–1208 (2019)
17. Sairam, B., Agrawal, A., Krishna, G., Sahu, S.P.: Automated vehicle parking slot detection system using deep learning. In: 2020 Fourth International Conference on Computing Methodologies and Communication (ICCMC), pp. 750–755 (2020)
18. Chen, L.-C., Sheu, R.-K., Peng, W.-Y., Wu, J.-H., Tseng, C.-H.: Video-based parking occupancy detection for smart control system. Appl. Sci. **10**, 1079 (2020)
19. Guo, X., Li, Y., Ling, H.: LIME: low-light image enhancement via illumination map estimation. IEEE Trans. Image Process. **26**, 982–993 (2017)
20. Deng, L., Yu, D., Platt, J.: Scalable stacking and learning for building deep architectures. In: 2012 IEEE International Conference on Acoustics, Speech and Signal Processing (ICASSP), pp. 2133–2136 (2012)
21. Zhao, K., Wiliem, A., Lovell, B.: Kernelised orthonormal random projection on grassmann manifolds with applications to action and gait-based gender recognition. In: 2015 IEEE International Conference on Identity, Security and Behavior Analysis (ISBA), pp. 1–6 (2015).
22. Chang, J.-M.: Classification on the Grassmannians: Theory and Applications. ProQuest (2008)
23. Gu, M., Eisenstat, S.: Efficient algorithms for computing a strong rank-revealing QR factorization. SIAM J. Sci. Comput. **17**, 848–869 (1996)
24. Dereniowski, D., Kubale, M.: Cholesky factorization of matrices in parallel and ranking of graphs. In: Parallel Processing and Applied Mathematics, pp. 985–992. Springer, Heidelberg (2003).

A Development Framework for Automated Facial Expression Recognition Systems

Bacha Rehman[1(✉)], Wee Hong Ong[2], and Trung Dung Ngo[3]

[1] Department of Computer Science, Namal Institute, Mianwali, Pakistan
bacha.rehman@namal.edu.pk

[2] Faculty of Science, Universiti Brunei Darussalam, Bandar Seri Begawan, Brunei Darussalam
weehong.ong@ubd.edu.bn

[3] The More-Than-One Robotics Laboratory, University of Prince Edward Island, Charlottetown, Canada
dungnt@ieee.org

Abstract. Automated facial expression recognition (AFER) has become an important research area with several computer vision (CV) applications. A robust AFER system requires sufficient good quality training and testing data for development and evaluation of a robust AFER model. There exist a number of AFER datasets and an increasing number of research works in AFER. However, research works in AFER have not matured to a stage that there are openly available platforms or toolsets to implement the pipeline of AFER system development. New comers to the field are faced with various challenges such as 1) images in the datasets are messy or with low resolutions; 2) the data are not organized into separate training and testing data for fair evaluation; 3) majority of the datasets are very small leading to insufficiency for training a model; 4) some datasets do not provide important facial features, 5) it is unclear which dataset to start with, and 6) no development framework and methodologies to systematically implement and test new models. In this paper, we present a framework with complete source code and algorithms to: 1) detect faces and crop face images in a given dataset for AFER; 2) extract facial landmark features from the face images and store as landmark images; 3) split the dataset into training and testing sets and stored into two CSV files consisting filename, emotion, landmarks, and features vectors for each image in its respective set; 4) train and evaluate the features vectors of the dataset using a deep neural network (DNN) model as the baseline; 5) train and evaluate a baseline convolutional neural network (CNN) on the face cropped images; 6) demonstrate the trained model on live videos and images. This study also outlines the necessary steps involved in developing an AFER system. The framework can help researchers to use a dataset to develop AFER systems and further improve the framework and benchmark the results.

Keywords: Facial keypoints · Facial expressions · Emotions recognition · Deep neural network · Convolutional neural network · Facial expressions dataset

W. S. H. Suhaili et al. (Eds.): CIIS 2020, AISC 1321, pp. 160–171, 2021.
https://doi.org/10.1007/978-3-030-68133-3_16

1 Introduction

Facial expressions are vital for affect signaling systems. It is the natural source of cues regarding a person's emotional state. Significant research has been recently done in the area of AFER. Several datasets have been created as well as different methods are applied to develop state of the art AFER systems [1]. Different machine learning models have been developed for AFER systems including convolutional neural network (CNN), deep neural network (DNN), support vector machine (SVM), local binary patterns (LBP), hybrid CNN + recurrent neural network (RNN), hybrid principal component analysis (PCA) + Latent Dirichlet allocation (LDA), and support vector regression (SVR) based action units (AU) intensity [1].

In this paper, we have utilized CNN and DNN based deep learning models as part of the framework provided in this paper. It is because the CNN models are one of the most popular deep learning models used for images and face related research. However, sometimes the CNN models fail to be adapted for real-world applications due to its robustness mainly caused by insufficient training data, simplified architecture of a particular CNN and the limited variance and diversity in the training data. Therefore, it is essential that researchers should know about alternative deep learning model, i.e. DNN in this paper. A DNN based baseline model is also provided as part of the development environment in this work so that researchers should adapt and use features based deep learning methods.

Facial expression datasets are essential to train machine learning algorithms for an AFER system. The accuracy of the training outcome is highly affected by the size and diversity in a dataset. There are many relevant facial images datasets [2], however, a few of them are conveniently useful to develop an AFER system. Finding and selecting suitable dataset is a challenging task especially for researchers new in the AFER domain. There are several challenges while selecting a database to train and test a model for AFER system:

- Some datasets are messy and have inaccurate image labels.
- Some datasets do not provide the facial landmark features useful for the development of AFER models.
- Several datasets consist of only a few hundred images and is not suitable to train a model especially using deep learning.
- Many datasets consist of only images for training without pre-defined sets of training and testing data to fairly evaluate and compare the trained model.
- The datasets usually do not provide sufficient details and tools to develop AFER systems with the datasets.

These limitations lead us to provide a development framework including DNN and CNN models, along with the suggestion of a posed expressions dataset that can also be used as spontaneous expressions dataset. This framework and the suggested dataset can suffice to allow AFER researchers to quickly develop a baseline AFER system and for further analysis and development. This paper explains the method adapted to create an organized AFER dataset from the extended Cohn-Kanade (CK+) dataset. The resulting dataset has eight facial expressions including 'Neutral', 'Angry', 'Disgust',

'Fear', 'Happy', 'Sad', 'Surprise', and 'Contempt'. The methodology behind the proposed framework has been explained in Fig. 2, while the proposed deep learning models have been elaborated in Sect. 5. The proposed framework includes:

1. Source code for image preprocessing, generate 68 facial landmarks, calculate distance vectors among every facial point and to save cropped images, landmark images and generate CSV file.
2. Source code to build, train, evaluate and save the baseline DNN and CNN models on the data provided in the CSV files and generate graphs for loss and confusion matrix.
3. Source code to apply the trained models on unseen images and videos including live video input.

The key contributions of this study are as follows:

- Provide a comprehensive development framework and source code for researchers to understand, implement and evaluate AFER systems.
- Describe the development pipeline including dataset preparation based on a suggested dataset.
- Provide details of how to expand the dataset and use it for both posed and spontaneous purposes.

2 Related Work

Several face related datasets are available for researchers to conduct experiments on AFER [1], face recognition, and facial attribute analysis [3]. However, each of them was created for different purposes and they lack the details and tools to ease its usage. It is a difficult task for a new researcher group in the field to find a suitable dataset to develop AFER models. The Japanese Female Facial Expression (JAFFE) [4], Maja-Michel-Ioannis (MMI) [5], facial expression recognition (FER2013) [6], the Karolinska directed emotional faces (KDEF) [7], and the Yale [8] are well known facial expressions databases available for research purposes. The main limitations of these datasets are that they did not provide any framework or tool to use effectively use the dataset and the datasets are not easily extendable. Furthermore, the datasets are either messy or have very limited number of images which is not suitable to train state of the art deep learning models, i.e. CNN Models.

A comprehensive dataset was first provided by Cohn-Kanade (CK) [9]. This dataset was created in the form of images sequences. Each images sequence was labeled using facial action coding system FACS [10] for the corresponding facial expression. The main drawback of this dataset was that the images sequences were unverified against the actual facial expressions they comprehended. Another drawback was the lack of intensity labels, which leaded this dataset to be extended to CK+ [11]. More samples of spontaneous expressions were added to CK+. The total number of images sequences in CK+ is 593, in which 327 images sequences have discrete facial expression labels. Images resolution of CK+ is 640×490. Table 1 summarizes the description and limitations of all these datasets.

Table 1. Overview of available AFER datasets (Image-Based)

Dataset	Images/videos	Limitations
FER 2013 [6]	28709 – training 3589 – test/valid each	The dataset is for competition hence very messy Images resolution is low Some emotion images reveal false facial expressions
CK [9]	2000 image sequences	Only image sequences Images sequences were unverified against the actual facial expressions
CK+ [11]	593 image sequences	Only image sequences 327 out of 593 image sequences are annotated Provided for spontaneous emotion analysis
KDEF [7]	490 images	Limited and small dataset Cropped images provided
JAFFE [4]	213 images	Limited and small dataset Only 213 images provided
MMI [5]	1500 images and image sequences	Included non-frontal faces Facial expressions added later,
Yale [8]	165 images	Small dataset not suitable for training a model

In development framework provided in this paper, we have used MTCNN [12] for face detection within the image frames because of its superior performance [13, 14] in comparison to Haar based algorithm [15] for face detection. For the facial landmark [3] extraction, we have used the DLIB library [16]. The experiments data, environments and settings were kept consistent for all the experiments in this work for uniform analysis and fair evaluation. The following deep learning models along with its given name and references were used to compare accuracy results with the proposed deep learning models in this work.

1. CNN-1 (Mini Xception) [17]
2. CNN-2 (Deep vs Shallow) [18]
3. CNN-3 (Deep CNN) [19]

3 Dataset Preparation

The CK+ dataset [11] was first divided into separate training and testing sets at a ratio of around 2:1. We derived each of the facial expressions from the available 593 sequences of the CK+ dataset. There are only 327 sequences which were annotated by the creator of CK+ dataset. We manually annotated the remaining 266 images sequences by thoroughly examining the nature of expression. There are 8 facial expressions including 'Neutral', 'Angry', 'Disgust', 'Fear', 'Happy', 'Sad', 'Surprise', and 'Contempt'. The images for each expression are organized into their respective folder.

In the CK+ dataset image sequences, the intensity of facial expression gradually increases frame-wise from Neutral towards the annotated facial expression. Keeping this in mind, the Training and Testing set have been created by taking images from the image-sequence. For instance, 1st and 2nd images in each sequence were taken as Neutral facial expression for the Training set, because each image sequence starts with a neutral facial expression. As the last two images in the sequence have the extreme visibility for the annotated facial expression, i.e. peak expression, therefore, the last two images were considered as the samples of the corresponding expression for the training set. For the testing set, the 3rd image in each sequence was taken as Neutral expression. For the other expressions, the visibility was determined from visual inspection. The search for the visibility initiated from the middle image in the images sequence until the 3rd last image of the sequence. The image with the most visible expression was taken as a sample for the test set. It was observed that the expression visibility was quite obvious for the 5th or 6th position from the last image in each sequence.

Table 2. Images details of the Dataset provided

Expression	Sub-folder	Training set	Testing set
Neutral	0	1174	586
Angry	1	112	56
Disgust	2	130	65
Fear	3	144	71
Happy	4	192	95
Sad	5	174	86
Surprise	6	192	96
Contempt	7	36	20

The total number of images in the resulting training set was 2154 while the testing set consisted of 1075 images as shown in Table 2. The resulting dataset can be extendable and more images from other datasets can be added into any facial expression to increase the number of training or testing samples. The workflow created an improved dataset with original facial expression images (640 × 490), cropped images (200 × 200), and facial landmark images (200 × 200 with 68 facial points). Each image was cropped around the face area detected using Multi-task Convolutional Neural Networks (MTCNN) [12]. The facial landmark [3] based images consist of 68 facial points obtained using Dlib library [16]. A CSV file was generated for the training and testing sets each. There were four fields in each CSV file, i.e. filename, emotion, landmarks, and landmarks distance vectors.

4 Dataset Pre-processing and Features Extraction

Once the images have been selected and placed into their respected folders, the images were processed and features were extracted so that the dataset can be conveniently utilized for any DNN or CNN models training. The process consists of mainly four tasks, i.e. face detection, face area cropping, facial landmarks distance vectors calculation, and save the cropped images, landmarked images and features vectors.

Once the face is detected through MTCNN, the next step is to calculate the 68 facial landmarks. Dlib library [16] has been used for calculating the facial landmarks. The face box returned by MTCNN is in rectangular shape. However the Dlib library requires the face box in a square shape. The forehead area of the face does not contribute much in the facial expression and determination of the facial landmarks. Therefore, the forehead area in the face box was discarded. The Eqs. (1) to (3) were used to discard the forehead area of the rectangular face box. The delta (Δh) is the number of row pixels to be cropped from the original face rectangle; h_o and w_o are the original height and width of the face rectangles; h_c is the cropped rectangle height; and x_o and y_o are the coordinates of the upper left corner of the original face rectangle. The region of interest (ROI) extracted from the original image i is a square face box suitable for the facial landmarks extraction.

$$\Delta h = h_o - w_o \tag{1}$$

$$h_c = h_o - \Delta h \tag{2}$$

$$ROI = i[y_o + h_c : y_o + h_o, x_o : x_o + w_o] \tag{3}$$

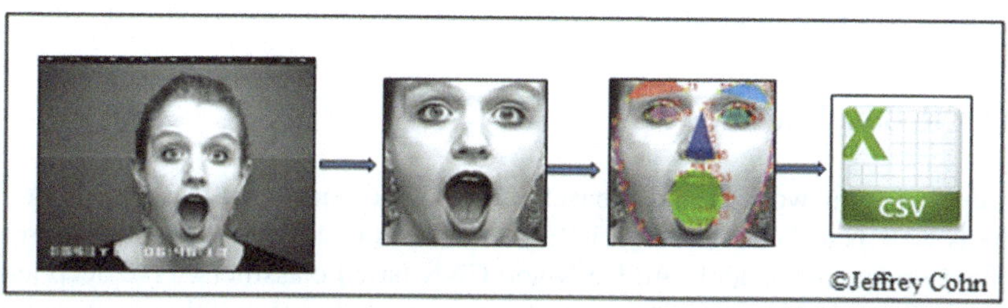

Fig. 1. Pre-processing and feature extraction for single image.

The Euclidean distance has been used to calculate the distance between all pairs of facial landmark points. Figure 1 depicts the pre-processing and feature extraction activities for a single image as described above. Figure 2 describes the complete process to create the organized dataset that has been implemented in the proposed framework.

Dataset preparation and data pre-processing process

Data: Images from a public dataset
Result: Organized dataset with folders and *CSV* files
Create folders *Train_Original, Test_Original, Train_Crop, Test_Crop, Train_LM, Test_LM.*
Create sub-folders *0* to *7* (eight expressions) in each of the above folders.
Split the original dataset into train and testing sets by selecting the images from image sequences, as described in Section 3, into their respective sub-folders (expressions) in *Train_Original* and *Test_Original.*
Open two new *CSV* files, *Train.csv* and *Test.csv.*

```
for each folder f in "Train_Original, Test_Original" do
    for each image i in each sub-folder in f do
        1. Detect the face
        2. if face is detected then
            2.1 Use Equations (1) to (3) to obtain the face box
            2.2 Crop the face box
            2.3 Resize the face box to 200 x 200
            2.4 Save the cropped image in the corresponding sub-folder in
                Train_Crop or Test_Crop depending on f
            2.5 Extract facial landmarks
            2.6 Save landmarked images in the corresponding sub-folder in Train_LM
                or Test_LM depending on f
            2.7 Calculate distance among all facial landmarks
            2.8 Add a line with filename, sub-folder name as emotion, landmarks, and
                distance vectors in Train.csv or Test.csv depending on f
        end
    end
end
```

Close and save both *CSV* files

Fig. 2. The pipeline of dataset preparation, data processing and feature extraction.

5 AFER Baseline Models

The proposed framework includes baseline models to conveniently adapt in an AFER system as well as to benchmark AFER systems developed with the same dataset. The baseline deep leaning models are DNN and CNN based classifiers. Five deep learning AFER DNN models were developed and evaluated as listed in Table 3. Based on the accuracy results, the model number 5 in Table 3 has been selected and included in the framework. Softmax has been used as the activation function in the output layer. The structure of the proposed DNN model consists of 6424 input features, 5×1024 hidden layers, and an output layer with 8 neurons. Min-Max features normalization has been applied for features normalization within 0 to 1 range.

Apart from the DNN based model, a CNN based model has also been designed and included in the framework. Figure 3, shows the architecture of the baseline CNN model. The face images are resized to 48×48. The proposed CNN model consists of 4 convolutional layers of size 64, 128, 512 and 512 respectively. The kernel size for each convolutional layer were taken as (3×3), (5×5), (5×5), and (3×3) respectively. Each

Table 3. DNN deep learning model tested

#	DNN structure	Accuracy
1	Input, 512, 512, 512, 512, 512, Output	88.55
2	Input, 512, 512, 512, 512, 512, 512, 512, Output	88.17
3	Input, 1024, 512, 512, 512, 512, Output	88.45
4	Input, 1024, 512, 512, 512, 512, 512, 512, Output	88.36
5	Input, 1024, 1024, 1024, 1024, 1024, Output	89.58

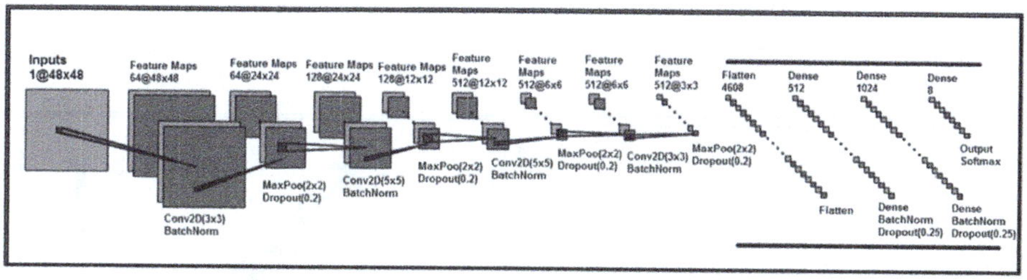

Fig. 3. Architecture of the CNN baseline model.

convolutional layer was followed by batch normalization, maximum pooling of pool size = (2 × 2), and a dropout layer of size 0.2 as shown in Fig. 3. After the fourth dropout layer, the features vectors were flatten using the Flatten layer. Two fully connected layers of size 512 and 1024 have been used after the Flatten layer. Each fully connected layer has been followed by a batch normalization layer and a dropout layer of size 0.25. Finally the output layer was placed after the final dropout layer.

Stochastic gradient descent (SGD) [20] was used as the model optimizer. Maximum number of epochs for the training process was set at 500. The learning rate of the SGD algorithm was set at 0.01. No stopping criteria were applied during the training process. We reduced learning rate during the training process with settings as factor = 0.5, patience = 50, and minimum learning rate = 0.0001. The batch size was set to 64.

6 Results and Discussion

Experiments were conducted using the two deep learning models described in Sect. 5. CPU computation has been used for both training and testing process. The results reported here are intended for benchmarking of other AFER systems developed using the suggested dataset. Figure 4 shows the loss graph and normalized confusion matrix (accuracy) for the training and testing data of the proposed DNN baseline model. Figure 5 shows the same information for the proposed CNN baseline model.

The benchmark accuracy achieved using the CNN based model was significantly higher than the result of the DNN based model. The accuracy achieved using the CNN model was 97.21% on the prepared testing set. However, it was observed that the DNN

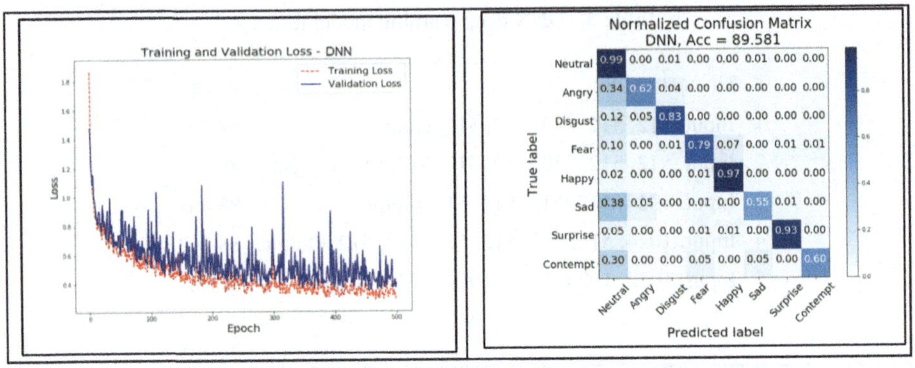

Fig. 4. Loss and normalized confusion matrix for DNN model

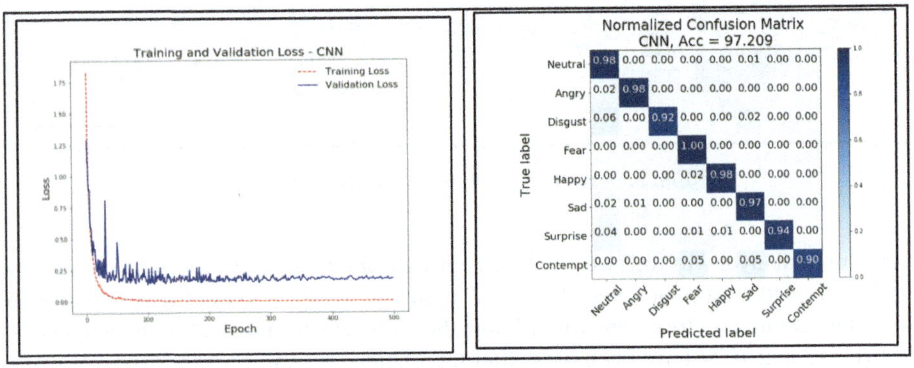

Fig. 5. Loss and normalized confusion matrix graph for CNN model

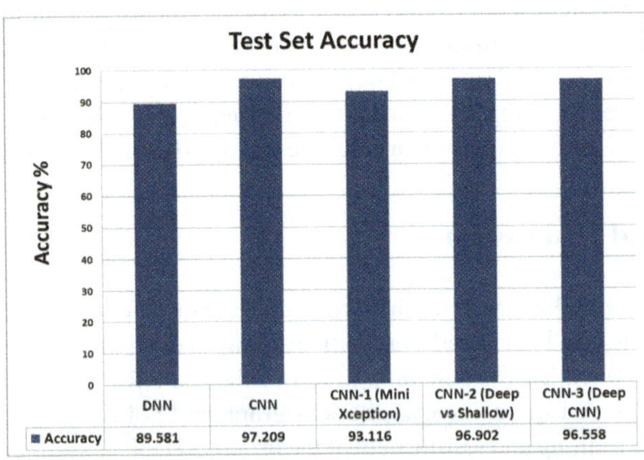

Fig. 6. Classification accuracy achieved by each deep learning model over the testing set

based model works quite robustly on the unseen data such as the live stream from webcam. DNN models are learned from extracted features. DNN models are less biased to the nature of the images and more subjects independent. The dataset generated is sufficient to train any features based DNN model. However, the CNN model can be made more robust by adding more data or using transfer learning [21]. A demo video is provided to show the performance of the baseline DNN model on live video of unseen subject [22]. Apart from the deep learning models proposed in this work, we also applied three deep learning models presented in [17, 18] and [19] on the dataset used in this work to compare the proposed models performance with the-state-of-the-art models. Figure 6 shows the classification accuracy for all the deep learning models for comparison.

7 Conclusion

A development framework consists of DNN and CNN based models are provided as a baseline framework for implementation and development of AFER systems. The framework can be used to develop, train, and evaluate any AFER system using deep learning models to recognize the eight common facial expressions. Experiment results have shown that the feature-based DNN baseline model exhibited robust real-time performance to detect correct facial expressions for the unseen subject. The suggested dataset can be used for both posed and spontaneous facial expression analysis. Lastly, the framework have in consideration to allow easy expansion of the generated dataset with more data to be added in order to train more robust CNN based models. The proposed framework can ease the learning curve of new researchers in developing their own AFER systems. The source code files for the development framework are provided on GitLab [23]. These source codes were used to generate CSV files, develop/train/evaluate deep learning models, and finally demonstrate the trained models over unseen images and live videos.

References

1. Mehta, D., Siddiqui, M.F.H., Javaid, A.Y.: Facial emotion recognition: a survey and real-world user experiences in mixed reality. Sensors (Switzerland) **18**, 1–24 (2018). https://doi.org/10.3390/s18020416
2. https://www.face-rec.org/databases/
3. Bulat, A., Tzimiropoulos, G.: How far are we from solving the 2D & 3D face alignment problem? (and a dataset of 230,000 3D facial landmarks). In: Proceedings of the IEEE International Conference on Computer Vision, pp. 1021–1030 (2017). https://doi.org/10.1109/ICCV.2017.116
4. Lyons, M.J., Shigeru, A., Miyuki, K., Jiro, G., Budynek, J.: The Japanese female facial expression (JAFFE) database. In: Proceedings of Third International Conference on Automatic Face and Gesture Recognition, pp. 14–16 (1998)
5. Pantic, M., Valstar, M., Rademaker, R., Maat, L.: Web-based database for facial expression analysis. In: IEEE International Conference Multimedia and Expo, pp. 5–8 (2005). https://doi.org/10.1109/ICME.2005.1521424

6. Goodfellow, I.J., et al.: Challenges in representation learning: a report on three machine learning contests. Neural Netw. **64**, 59–63 (2015). https://doi.org/10.1016/j.neunet.2014.09.005
7. Goeleven, E., De Raedt, R., Leyman, L., Verschuere, B.: The Karolinska directed emotional faces: a validation study. Cogn. Emot. **22**, 1094–1118 (2008). https://doi.org/10.1080/02699930701626582
8. Georghiades, A., Belhumeur, P., Kriegman, D.: From few to many: generative models for recognition under variable pose and illumination. IEEE Trans. Pattern Anal. Mach. Intell. **23**, 643–660 (2000)
9. Kanade, T., Cohn, J.F., Tian, Y.: Comprehensive database for facial expression analysis. In: 4th IEEE International Conference on Automatic Face and Gesture Recognition, FG, pp. 46–53 (2000). https://doi.org/10.1109/AFGR.2000.840611
10. Hamm, J., Kohler, C.G., Gur, R.C., Verma, R.: Automated facial action coding system for dynamic analysis of facial expressions in neuropsychiatric disorders. J. Neurosci. Methods. **200**, 237–256 (2012). https://doi.org/10.1016/j.jneumeth.2011.06.023.Automated
11. Lucey, P., Cohn, J.F., Kanade, T., et al.: The Extended Cohn-Kanade Dataset (CK+): a complete dataset for action unit and emotion-specified expression. In: 2010 IEEE Computer Society Conference on Computer Vision and Pattern Recognition Workshops (CVPRW), pp. 94–101 (2010). https://doi.org/10.18632/aginh.101501
12. Zhang, K., Zhang, Z., Li, Z., Member, S., Qiao, Y.: Joint face detection and alignment using multi-task cascaded convolutional networks. IEEE Signal Process. Lett. **23**, 1499–1503 (2016). https://doi.org/10.1109/LSP.2016.2603342
13. Rehman, B., Ong, W.H., Hong, A.T.C.: Using margin-based region of interest technique with multi-task convolutional neural network and template matching for robust face detection and tracking system. In: Proceedings of 2nd International Conference on Imaging, Signal Processing and Communication (ICISPC), pp. 14–18 (2018)
14. Rehman, B., Ong, W.H., Tan, A.C.H., Ngo, T.D.: Face detection and tracking using hybrid margin-based ROI techniques. Vis. Comput., 1–15 (2019). https://doi.org/10.1007/s00371-019-01649-y
15. Rehman, B., Ong, W.H., Hong, A.T.C.: Hybrid model with margin-based real-time face detection and tracking. In: The 11th Multi-disciplinary International Workshop on Artificial Intelligence (MIWAI). Lecture Notes in Computer Science, pp. 360–369. Springer, Cham (2017). https://doi.org/10.1007/978-3-319-49397-8
16. Baltrusaitis, T., Robinson, P., Morency, L.P.: OpenFace: an open source facial behavior analysis toolkit. In: IEEE Winter Conference on Applications of Computer Vision, WACV, p. 10 (2016). https://doi.org/10.1109/WACV.2016.7477553
17. Arriaga, O., Valdenegro-Toro, M., Plöger, P.: Real-time convolutional neural networks for emotion and gender classification. In: Proceedings of the 27th European Symposium on Artificial Neural Networks, Computational Intelligence and Machine Learning (ESANN 2019), Belgium, 24–26 April 2017 (2017)
18. Alizadeh, S., Fazel, A.: Convolutional neural networks for facial expression recognition. arXiv Prepr. arXiv:1704.06756 (2017)
19. Yu, Z., Zhang, C.: Image based static facial expression recognition with multiple deep network learning. In: Proceedings of the 2015 ACM on International Conference on Multimodal Interaction, pp. 435–442 (2015). https://doi.org/10.1145/2818346.2830595
20. Babenko, B., Yang, M.H., Belongie, S.: A family of online boosting algorithms. In: Proceedings of the IEEE International Conference on Computer Vision Workshops, ICCV Workshops, pp. 1346–1353 (2009). https://doi.org/10.1109/ICCVW.2009.5457453
21. Ng, H.-W., et al.: Deep learning for emotion recognition on small datasets using transfer learning. In: Proceedings of the 2015 ACM on International Conference on Multimodal Interaction, pp. 443–449 (2015). https://doi.org/10.1145/2818346.2830593

22. AI Lab: Multimodal Human Intention Perception For Human Robot Interaction. Ailab.space. https://ailab.space/projects/multimodal-human-intention-perception/#visual-facial-expression-recognition. Accessed 4 Sept 2020
23. Rehman, B., Ong, W.H., Ngo, T.D.: A development framework for Automated Facial Expression Recognition Systems, GitLab project (2020). https://gitlab.com/ailab.space/bacha-afer-dev-framework

Use Case of HPVT in Counting Stars: A Vision-Based Approach of Monitoring Natural Events

Muizz Kasim[1], Ken T. Murata[2], Najibudin Midun[1], Takamichi Mizuhara[3], Praphan Pavarangkoon[2], and Somnuk Phon-Amnuaisuk[1](✉)

[1] Centre for Innovative Engineering, Universiti Teknologi Brunei, Gadong, Brunei
{B20161128,B20161106}@student.utb.edu.bn,
somnuk.phonamnuaisuk@utb.edu.bn

[2] National Institute of Information and Communications Technology, Tokyo, Japan
{ken.murata,praphan}@nict.go.jp

[3] CLEALINK TECHNOLOGY Co., Ltd., Kyoto, Japan
mizuhara@clealink.jp

Abstract. Counting stars could reveal the level of a cloudy sky at night, which provides important climate information. The monitoring of natural events, such as counting stars, can pose as a highly demanding task which requires an immense amount of time and dedication to be completed by a human observer. By using rich visual data from a video transmission system, namely HpVT, as the input, a specialized monitoring program utilizing computer vision techniques was developed to count stars under various night sky conditions. We present the design of a star counting framework and describe the counting process employed. We perform testing on the star counting framework using our own data set. The data set consists of 154 video footages (of 24-h period each) contain footages of the night sky which were recorded from various locations in different countries. This work shows a proof of concept that the number of stars observed can correlate well with climate information. We also discuss some of the issues faced and make recommendations for future projects.

Keywords: Natural event monitoring · Computer vision · HpVT · Raspberry Pi · Counting stars

1 Introduction

Components in our ecosystem seem to relate with deep relationships. Observation of events in our ecosystem has revealed many interesting facts about activities of living organisms and non-living components. Monitoring events in nature which may appear trivial at the first glance may reveal deeper understanding about our ecosystem. Information regarding the number of stars might reveal interesting patterns of weather.

In this paper, we count the numbers of stars at night-time. In order to record visual information throughout the night for over a long period of time (i.e., in the time scale

W. S. H. Suhaili et al. (Eds.): CIIS 2020, AISC 1321, pp. 172–181, 2021.
https://doi.org/10.1007/978-3-030-68133-3_17

of months or years), the data acquisition system is an important engineering task. We leverage on our own *High-performance Video Transmission* (HpVT) platform for the data acquisition task. The HpVT platform [1] allows operators at the data center to control visual data acquisition from a camera connected to *Raspberry-Pi* (RPi) computers at remote sites.

The work in this paper is divided into several sections. Related past works are covered in Sect. 2. Section 3 describes the implementation of our counting program or framework. We disclose our results in Sect. 4. Section 5 is our discussion section. And the conclusion is in Sect. 6.

2 Related Works

The monitoring of natural events can bring about many benefits from its application. Knowledge extraction of the events happening in our surrounding environments could reveal interesting relations and interactions among components in the environment. Many works have been done in previous years which explore this branch of science for a variety of different reasons or purposes. One such purpose is for the monitoring of natural hazards such as in works done by [2, 3], where the monitoring of forest fires was achieved using *unmanned aerial vehicle* (UAV). In a different light, studies on the monitoring of natural resources such as in [4, 5] have also been conducted, where researchers used various imaging technology to estimate and map the heights of forest canopies. In another example, [6, 7], monitoring of the skies and clouds using ground-based sky imagery technique was used in solar forecasting for *direct normal irradiance* (DNI) of the sun.

2.1 Application of Natural Event Detection and Analysis

The use of image processing techniques in the task of monitoring natural events has been increasing in recent years. Studies conducted in [8–12] demonstrate this clearly.

Many research studies have explored the domain of stars in the past several decades [13]. Algorithms or techniques in star identification and pattern recognition have been developed previously such as in [14–17]. However, these algorithms and techniques were developed for use in star trackers, which are devices typically used to determine the altitude or orientation of a spacecraft.

Not a lot of attention has been given on the development of star counting method or technique which are based on ground sky imagery. This also seems to be new, as recent articles have not made any mentions of any other projects which are like our own. The closest was in [18, 19], where firefly population data was collected in the USA as a response to the dwindling number of firefly population in that area. Although at first glance this work seems related to our own, our objectives differ and our method of achieving the objective also differs. Which, in our case, we use a computer vision approach to reach our objective.

2.2 Background of HpVT

The term HpVT stands for *High performance Video Transmission.* It is a protocol which transmits large amounts of visual data (video/picture) over the internet at high rates of transfer, in real-time. The HpVT was originally developed by a team of researchers from NICT in Tokyo, Japan [1]. It was made for the purpose of real-time outdoor monitoring of proposed locations (e.g., riverbanks, mountains, power plants, airports, volcanoes, etc.). To achieve this objective, the research team developed a video transmission system which incorporates the HpVT protocol by specifically designing it to work in conjunction with RPi computers, which are utilized heavily in the implementation of the system. Through the use of visual Internet of Things (IoT) devices, the system is able to collect large visual data of proposed outdoor location(s) and transfer them into a cloud network where it can be viewed and (or) manipulated from a web application (also developed by the same team). Data produced by the HpVT can then be accessed online (i.e., downloaded/streamed) from the web application where it can then be used in the task of monitoring environments remotely such as monitoring for volcanic activity, cloud patterns or counting stars.

3 Materials and Methods

A star counting program or framework was developed for this study. This program makes use of OpenCV library to detect and perform counting of the stars. We use Orange (a data visualization tool) to plot and analyze the data. In addition, we also perform testing on our proposed framework using a data set produced by our HpVT platform to demonstrate our framework design.

Data Collection Sites: Our HpVT platform collects data from three different recording sites. Data collected from these sites are used as test data for our star counting framework. Each recording site was chosen due to their natural disposition to harsh climate. Figure 1 provides sample images taken from each of these recording sites. Details regarding these recording sites (named A – C) are described below:

Recording Site A: located at Athabasca, Canada. Footages were captured using high sensitivity camera in Full HD (1080p) resolution. All footages were recorded in the year 2019.

Climate Description: Footages are susceptible to snowstorms and occasional auroras which appear near the horizon. Aside from that, the location also experiences moderate amount of dense and partially dense cloud formations. All of these factors can obstruct the appearance of stars at night.

Recording Site B: located in Iceland. Footages were captured using high sensitivity camera in Full HD (1080p) resolution. Footages were recorded between December 2018 to January 2019.

Climate Description: Footages are also susceptible to snowstorms. Similarly, auroras may also appear, which covers the whole sky due to Iceland being a polar region. The location experiences dense cloud formations most often. All of these factors can obstruct the appearance of stars at night.

Recording Site C: located near Mount Asama in Japan. Footages were captured using very high sensitivity camera in HD (720p) and SD (480p) resolutions. Footages were recorded between October 2017 to April 2018.

Climate Description: Mount Asama is an active volcano complex located in the island of Honshu, Japan. As such, footages are susceptible to volcano smoke. Storms and fog (due to rainfall) may also occur in the area. All of these factors can obstruct the appearance of stars at night time.

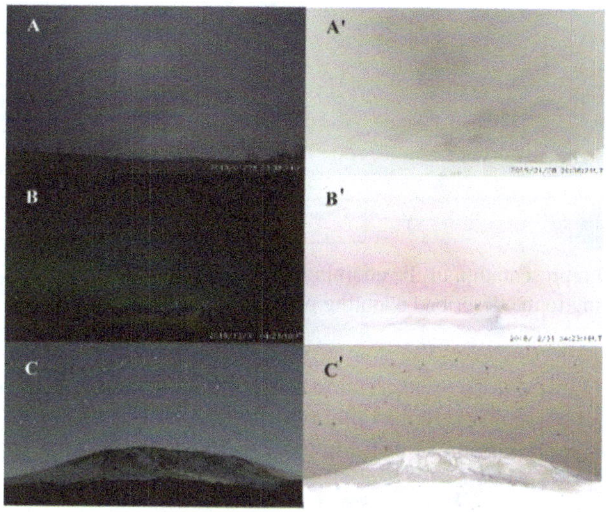

Fig. 1. Snapshot samples taken from recording sites A, B, and C respectively. Samples are shown in original colors (left) and reversed colors (right, for better contrast).

3.1 Nature of Test Data

The data set contains a total of 154 video footages spread across multiple months for all three recording sites. All video footages run at 30 FPS and cover a 24 h' time period which have been sped up akin to a time lapse, over a duration of 4.5 min. Each video footage therefore contains approximately a total of 8,100 frames. In total, there are approximately 1.2 million frames to be analyzed by the star counting program or framework.

Problems with Missing Data. Initially, our data set consist of 201 video footages. However, 47 of them either do not cover a 24 h' time period, or the video footage has experienced a malfunction. This was caused by hardware issues (equipment occasionally breaks down due to long exposures to harsh climate) which then leaves some of the hourly star count readings empty. This directly leads to the problem of missing data. As a result, only the remaining 154 video footages were used in testing.

3.2 Counting Process

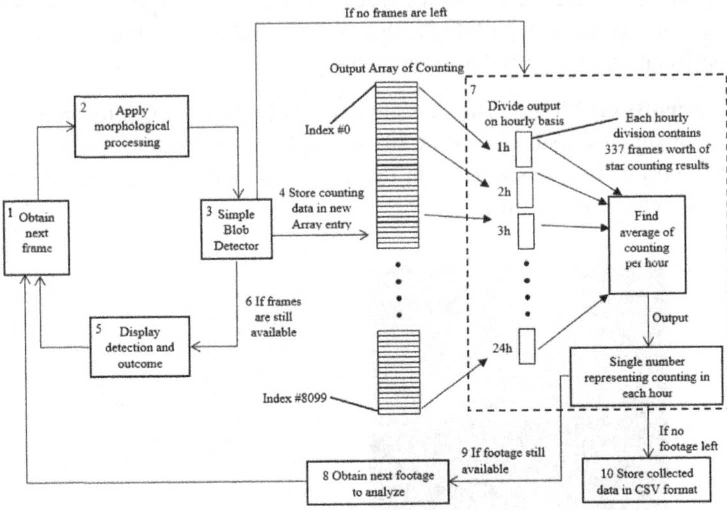

Fig. 2. Graphical representation of the counting process. Each functional block or condition are numbered according to the described counting process list.

Our star counting framework is designed to count the number of stars in a given footage. Figure 2 provides a graphical representation of our counting process for better understanding. A typical operation of our framework follows the counting process described in the list below:

(1) Obtain the next frame from input image.
(2) Apply morphological processing (if needed).

  i. For each frame extracted, several image processing techniques are applied to it to prepare it for the counting process. Techniques applied typically include masking (to remove unwanted backgrounds), thresholding (to segment or binarize image) and blurring (to remove noise). Any combination of these techniques is applied to each of the frames extracted.
  ii. It must be noted that not all of the footages undergo the same combination of morphological processes however, as this issue is highly subjective to the problems present in each input footage. For instance, a footage obtained at location A might require all three techniques to prepare it. While a footage obtained at location B might only require masking and blurring but not thresholding.

(3) Begin detection of stars using SimpleBlobDetector class.

  i. Obtain the output data of detection, i.e., total count of objects detected, together with its corresponding frame interval reading.

(4) Record the output data in an array.

    i. Count data of star in each frame is stored as a separate new entry in the array.

(5) Display the detection and counting outcome unto a display monitor for viewing.
(6) Repeat step 1 to 4 until no more frames are available.
(7) From the recorded output data in the array, average the counting data collected for every hourly interval in the video in order to obtain a single counting variable representing each passing hour in the video footage analyzed.
(8) Obtain the next video footage (from the recording sites) to be analyzed.
(9) Repeat step 1 to 7 until no more footages are left.
(10) Compile all output data into a CSV file format.

SimpleBlobDetector Parameters. The effectiveness of counting can be influenced by a number of parameters such as the area, convexity, inertia and circularity of blobs. The blob attributes of convexity, inertia and circularity were set to have a minimum value of 0 and a maximum of 1. Minimum accepted area of detected blob is set to 1 px. If the extracted frame was thresholded before the counting process is applied, then the filterByColor parameter attribute is enabled and its value is set to 255. Otherwise, it is disabled. Lastly, the minimum distance between blobs is set to 3 px. It should be noted however that these parameters are highly subjective to the conditions present in each recording site. As such, variations of the settings described above were used in locations where the effectiveness of counting was deemed to be inadequate.

4 Results

Figure 3 shows the star counting outcomes of our tests which are plotted against the time in hours. Counting figure is shown on an hourly basis which spans over a 24-h period. Each line graph plot has also been labelled according to which recording site they had been gathered from. For recording site C, the graph plot has been split into two parts (C I. and C II.) to avoid the over-cluttering of information.

Results from our testing show different average hourly counting recorded over each month. These differences are to be expected in the counting of stars as it is known that the number of stars in the night sky changes with seasons. This change is in fact due to the rotation and orbit of the earth around the sun. Since all recording sites were in the northern hemisphere, references on the seasons refer to the meteorological definitions of the four seasons of winter, spring, summer, and autumn.

Varying levels of mean star counting in each month also suggests different levels of obstructions of the night sky were present in each area. Obstructions may be caused by the presence of naturally occurring events such as volcano smoke, fog (due to rainfall), aurora, snowstorm, and by the formations of dense or partially dense clouds which appear throughout the night time; All of which can give an important climate information of particular local area or region. Information such as these may be useful when used in conjunction with other meteorological data which may also be used to further studies involving climate change.

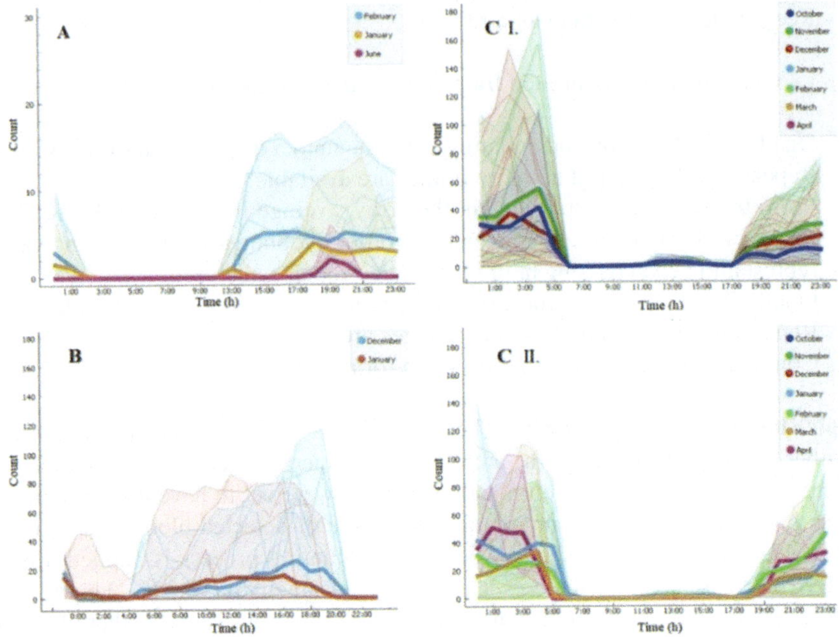

Fig. 3. Line graph plots of star counting against time for all recording sites; labelled A, B and C respectively. Each thin colored line represents a single day's counting in a particular month. Thick colored lines represent the mean counting for each month.

4.1 Potential Improvement Feature

The program can also be made to predict the motion of stars. This feature was not included in the implementation of the counting process described previously, but it can potentially improve its counting results and will be exploited in our future work.

Let W be the weightage coefficient and f be any measurable property of the object of interest (e.g. x and y positions, distance, size, intensity, etc.).

$$\sum_{i=1}^{n} W_i f_i \tag{1}$$

Equation (1) can be used as a way of recognizing the same entity of star across multiple frames. Properties of a certain star at frame k are subtracted with the properties of all stars data entry at frame k−1. In doing so, a list of outcomes of Eq. (1) is collected, which are then sorted in an ascending order. The first entry in the list of outcomes should be a positive number with a possible minimum outcome of 0. The interpretation of this number is that the outcome which is closest to the value of 0 is the most likely candidate to belong to the same entity as the star which has been previously selected. Additionally, statistical data of the star's properties can also be obtained if this process is repeated for a number of frames. Figure 4 below shows a graphical representation of this concept to give a better understanding to the reader (Table 1).

With the normalization of the output, the tracking accuracy should be improved. The major limitation of this process, however, is that the statistical data of the stars' properties

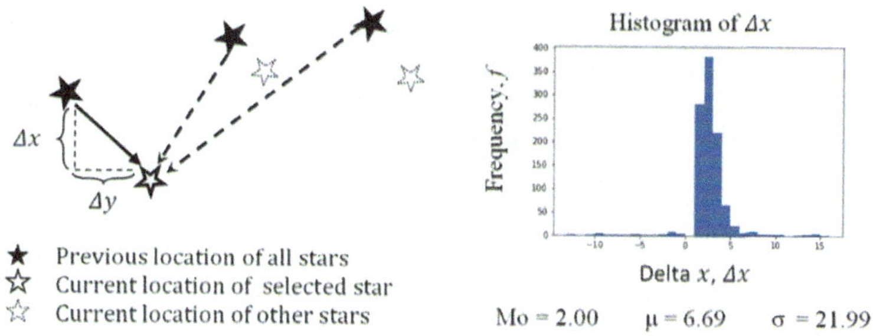

Fig. 4. Concept of object tracking with Eq. (1), along with an example of obtainable statistical data of the star's properties. Output closest to 0 is represented with an arrow of solid black line. Whereas all other outputs are represented as arrows of dotted lines.

Table 1. Example of outputs comparing unhomogenized (Output A) and normalized (Output B) outcomes of Eq. (1).

Property	Value	Weightage	Mean	Standard dev.	Output A	Output B
Delta x	4.44	0.45	2.13	37.81	2.00	0.449
Delta y	7.20	0.45	3.50	47.22	3.24	0.449
Delta s	0.40	0.10	0.02	1.73	0.40	0.098
WSF total	–	1.00	–	–	5.64	0.996

are required for the normalization calculations. This implies that the program must be executed at least once before any refinements to the algorithm can be implemented.

5 Discussion

In this paper, we have tested our star counting framework using our own data set. Unfortunately, however, the data set we use is asynchronous in nature. This is actualized in two ways: (1) asynchrony in location, and (2) asynchrony in time. Regarding the former, the footages were not captured within the same region or location (i.e., within the same country or locality). And regarding the latter, they were not captured during the same time frame (in years and months). Because of this, there was ultimately not enough synchronous data from which a strong claim regarding the weather patterns can be made. For that reason, we would argue that future projects should gather more synchronous data. So that, with enough data gathered, interesting patterns in the data may be revealed.

The vision-based approach of our framework also provides several advantages to the user. Among them include the lack of need for a manual observer. Practically, employing someone to perform the counting of stars throughout the night is realistically impossible, as this action requires immense amount of time and dedication to perform. Not only that, doing this might also put said observer at a great risk of harm, as he will be exposed to potentially dangerous climate conditions or the wildlife. In contrast, our method of counting allows for the remote monitoring of stars' population in locations which are otherwise dangerous, inaccessible, or difficult to reach.

That said, there were several issues faced regarding the timing nature of some of our data set. In earlier (pre 2018) renditions of the HpVT data, there were no inclusion of a timestamp display in the video footages themselves. This made it difficult to determine the actual time range within the video. The judgement of time had to be made manually by observing the rising and setting of the sun (or sunlight) within the selected footage. Although this issue might seem trivial, in some locations, especially in places where thick fog is common (e.g. near volcanoes), the manual judgement of time can become a real challenge which introduces unnecessary complications to the problem. This knowledge of timing is necessary to know the exact time range when the nighttime begins and ends.

6 Conclusion

In this paper, we have proposed a natural event counting program designed for counting stars which implements the SimpleBlobDetector algorithm from OpenCV library. Multiple tests were performed on various locations originating from different countries which shows the robustness of the program despite differing environmental conditions. The current data set presented in this paper are inadequate to make a strong claim in deciphering any interesting patterns in data, in part due to the asynchronous nature of the data used. Nevertheless, a framework for the monitoring of stars' population has been laid, which is ready for use in future projects. Other lessons can also be taken from this study which can help improve future studies of similar nature as the one covered in this paper.

Acknowledgments. This research work was a collaborative work between Center for Innovative Engineering (CIE) organization from Universiti Teknologi Brunei (UTB), and the National Institute of Information and Communications Technology (NICT) institution from Tokyo, Japan. Primary funding, research dataset, and technical support for the work was provided by the NICT institution, along with other involving cooperating parties such as the CLEALINK Technology Co. Ltd.

References

1. Murata, K.T., Mizuhara, T., Pavarangkoon, P., Yamamoto, K., Muranaga, K., Aoki, T.: Design and development of real-time video transmission system using visual IoT device. In: 12th International Conference, MIWAI 2018, Hanoi, Vietnam (2018)
2. Merino, L., Caballero, F., Martínez-de-Dios, J., Maza, I., Ollero, A.: An unmanned aircraft system for automatic forest fire monitoring and measurement. J. Intell. Rob. Syst. **65**(1–4), 533–548 (2011)
3. Casbeer, D., Sai-Ming, L., Beard, R., Mehra, R., McLain, T.: Forest fire monitoring with multiple small UAVs. In: Proceedings of the 2005 American Control Conference, Evanston, Illinois, pp. 3530–3535 (2005)
4. Birdal, A., Avdan, U., Türk, T.: Estimating tree heights with images from an unmanned aerial vehicle. Geomat. Nat. Haz. Risk **8**(2), 1144–1156 (2017)

5. Berni, J., Zarco-Tejada, P., Sepulcre-Cantó, G., Fereres, E., Villalobos, F.: Mapping canopy conductance and CWSI in olive orchards using high resolution thermal remote sensing imagery. Remote Sens. Environ. **113**(11), 2380–2388 (2009)
6. Chi, W.C., Urquhart, B., Lave, M., Dominguez, A., Kleissl, J., Shields, J., Washom, B.: Intra-hour forecasting with a total sky imager at the UC San Diego solar energy testbed. Sol. Energy **85**(11), 2881–2893 (2011)
7. Chi, W.C., Belongie, S., Kleissl, J.: Cloud motion and stability estimation for intra-hour solar forecasting. Sol. Energy **115**, 645–655 (2015)
8. Fei, Y., Yifan, H., Xin, C., En, C.: A biological sensor system using computer vision for water quality monitoring. IEEE Access **6**, 61535–61546 (2018)
9. Chi, Y., Zhixiang, L., Youmin, Z.: UAV-based forest fire detection and tracking using image processing techniques. In: 2015 International Conference on Unmanned Aircraft Systems (ICUAS), Denver, CO, USA. IEEE (2015)
10. Chi, Y., Zhixiang, L., Youmin, Z.: Vision-based forest fire detection in aerial images for firefighting using UAVs. In: 2016 International Conference on Unmanned Aircraft Systems (ICUAS), Arlington, VA, USA. IEEE (2016)
11. Simelli, I., Tsagaris, A.: The use of unmanned aerial systems (UAS) in agriculture. In: International Conference on Information and Communication Technologies in Agriculture, Food and Environment (HAICTA 2015), Kavala, Greece (2015)
12. Wood-Bradley, P., Zapata, J., Pye, J.: Cloud tracking with optical flow for short-term solar forecasting. In: Proceedings of the 50th Annual Conference of the Australian Solar Energy Society, Melbourne (2012)
13. Spratling, B.B., Mortari, D.: A survey on star identification algorithms. Algorithms **2**, 93–107 (2009)
14. Samann, M., Mortari, D., Junkins, J.: Recursive mode star identification algorithms. IEEE Trans. Aerosp. Electron. Syst. **41**(4), 1246–1254 (2015)
15. Padgett, C., Kreutz-Delgado, K.: A grid algorithm for autonomous star identification. IEEE Trans. Aerosp. Electron. Syst. **33**(1), 202–213 (1997)
16. Mortari, D.: Search-less algorithm for star pattern recognition. J. Astronaut. Sci. **45**(2), 179–194 (1997)
17. Hong, J., Dickerson, J.A.: Neural-network-based autonomous star identification algorithm. J. Guid. Control Dynam. **23**(4), 728–735 (2000)
18. Chow, A.T., Chong, J., Cook, M., White, D.: Vanishing fireflies: a citizen-science project promoting scientific inquiry and environmental stewardship. Sci. Educ. Civ. Engagem. **6**, 23–31 (2014)
19. White, D., Pargas, R., Chow, A., Chong, J., Cook, M., Tak, I.: The vanishing firefly project. In: IWGS 2014: Proceedings of the 5th ACM SIGSPATIAL International Workshop on GeoStreaming, pp. 85–92 (2014)

Investigating Haptic Feedback for Human Interactions in Virtual Environment

Amal Rafidah Haji Hamzah(✉) and Nurulhidayati Haji Mohd Sani

School of Computing and Informatics, Universiti Teknologi Brunei,
Jalan Tungku Link, Gadong, Bandar Seri Begawan BE1410, Brunei Darussalam
{m20190009,p20151003}@student.utb.edu.bn

Abstract. The sense of touch is essential for our daily life activities. It helps human beings to differentiate the characteristics of the objects in the environments. Same goes in Virtual Reality (VR), it not just replicating the real-world objects but the haptic sensation as well. Integrating electro-cutaneous device and haptic feedback system enables the simulation of touch sensation inside the virtual environment. In this project, the haptic glove prototype was built to produce touch feedback through vibration and implementing it to the virtual environment inside Unity3D. The haptic glove is controlled with Arduino and programmed in Unity3D to create a relationship between the haptic device and the Leap Motion Controller (LMC). The Leap Motion Controller detect and track the user's hands and emulate the virtual hands displayed in Unity3D. Different sizes of virtual cubes were created inside the Unity3D to evaluate the functionalities and the capabilities of the haptic glove. The collision of the virtual hand and the cubes shall trigger the vibration sensation to the user's fingertips. The experiments show that the addition of haptic device promotes better interactivity between the user and the virtual objects. In this work, the presence of virtual objects is perceived in the form of vibration. In future work, we shall design the efficient system to reach the three desired features of VR: immersion, interaction and presence.

Keywords: Haptic device · Arduino · Leap motion controller · Virtual environment · Unity3D

1 Introduction

Virtual Environment (VE) provides a safer way for users to experience the realism of real-life situations. The study of realism in VE has been around in the literature for the past decade [1]. Realistic VE creates a better user experience. Realism in terms of visual and auditory technologies have been distinctly achieved in the ongoing trends of VE that provides an immersive experience in the environment and are commonly implemented in every multimedia setting like in Virtual Reality (VR). VR is the instance of VE that resembles the real

W. S. H. Suhaili et al. (Eds.): CIIS 2020, AISC 1321, pp. 182–191, 2021.
https://doi.org/10.1007/978-3-030-68133-3_18

world [2] and may be able to form a synthetic sensation and produce immersive feelings to the users [3]. It could bring a user to a different world of dimension. Furthermore, the applications of VR have the potential in constructing complex simulations for training critical operation skills such as flight simulations.

Although this is the case, many of these experiences lack the immersive of the touch sensation. Researches have been investigating ways to resolve the issue and haptics technologies is one of the methods that enable this desire [3,4]. Haptic senses are essential to almost everyone in daily life to feel the objects and environments and we often experience these senses through human skins [5]. Moreover, human hands are one of the organs that used to manipulate things and hence we able to differentiate the characteristics of the object only by touch.

The additional of haptic device to the VEs may be beneficial for the users to experience a realistic simulation of force and tactile. Although having devices such as wheels and joystick to form a vibrotactile and proved to provide a good control in VR, Leap Motion Controller (LMC) may give a better alternative way to interact with the virtual objects. LMC often used for communication between human and the digital devices through hand gestures. LMC allows detection and tracking of the hands and fingertips movements within one meter radius.

In this work, we build a prototype haptic device that enhances touch experience in VE using vibrations. When users are touching and interacting with the virtual objects inside the VEs, they will experience the presence of the virtual objects and allows them to have a more natural object manipulation and interaction. The LMC detect and track the movement of the haptic glove, and if the virtual hand touches the virtual objects, corresponding motors attached to each finger on the physical glove will vibrate.

The outline of the paper is as follows. Related work is discussed in Sect. 2. The development of haptic glove will be discussed in Sect. 3. Section 4 shows the experiments of the device and discussion of the result is discussed in Sect. 5. Finally, main conclusions are summarized in Sect. 6.

2 Related Works

A haptic device allows user to sense the tactile and kinesthetic information when interacting with the environment, be it real or virtual [6]. It is a mechanical device that allow users to touch, feel and manipulate the virtual objects in VEs. The haptic devices are considered as input-output devices as it tracks user's physical manipulations (input) and provides a realistic touch sensation with on-screen events (outputs).

The concept of touch in the real and virtual world has been widely investigated in previous literature, it often comes in the form of a glove or vest. As for haptic glove, [3] classified the technology into three groups: traditional gloves, thimbles and exoskeletons.

The traditional glove is a fabric glove attached with flexible wires and mini motors that allows immersive interactions through flex and vibration sensors. This haptic glove can be used to identify touched objects in VE such as solid rocks or roaring waterfall [7]. The main disadvantage of using the glove is that to experience a total immersion, the glove must be perfectly fit to the hand of the wearer so that the motor is intact with the wearer's hands. Furthermore, flex sensors may cause discomfort during bending among wearer's joints.

Thimbles works by placing the motor sensors with the actuators to the user's fingertips to allow immersive, tactile interactions in VE [8] on that area. Thimbles may not only cause discomfort, but it also may be painful if it is worn for a long duration of time if it is not carefully designed.

The exoskeleton is often developed for training and rehabilitation purposes as it helps in assisting physical therapy for the rehabilitation process [9]. CyberGrasp [10] and VRgluv [11] are examples of the exoskeleton and popular in their early introduction to the market. These gloves were designed wirelessly to allow users to move freely in the virtual environment. With the combination of the precision hand tracking and force feedback technologies, it seamlessly transports the virtual object right to the hands.

3 Experimental Design

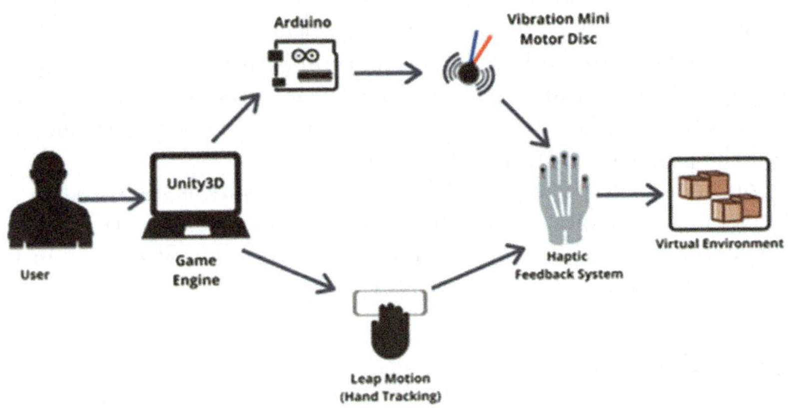

Fig. 1. Framework of haptic feedback system in Virtual Environment (VE)

Figure 1 shows the framework of the process involved in designing the Haptic feedback system in VE. User will first interact with the environment using Unity3D and while wearing the haptic glove. Leap motion will track the movement made by the glove and act accordingly to the environment. Once an object is touched, the system in Unity3D will then send feedback to the haptic glove stating that collision has occurred. The Arduino will send signal to the haptic glove to produce a vibration and hence notify user when an object has been touched. The communication of these devices is two-way direction.

The designing of the haptic feedback system is divided into two parts: (i) designing a prototype of haptic device; and (ii) designing the VEs in Unity3D.

The prototype for the haptic device only focuses on one stimulus; tactile and the output will be based on vibrotactile sensation. This haptic device was designed in the form of glove-like. As for virtual environment, the simple translation task was set up in order to test the capabilities and the functionalities of the haptic feedback.

3.1 Designing a Prototype Haptic Device

The haptic glove (see Fig. 2) was designed to generate the tactile simulation during interactions in VE. The Arduino Uno microcontroller contains a Pulse Width Modulation (PWM) which is incorporated with a mini vibrating motors produced actuators. The glove contains five of these vibrating motors (i.e. one on each finger) that have an adjustable frequency range between 0 and 350 Hz and rated voltage between 0 to 5 V. These motors are controlled by Arduino Uno board by programming the code with Arduino IDE. The haptic device has the features of low latency and supports the right hand.

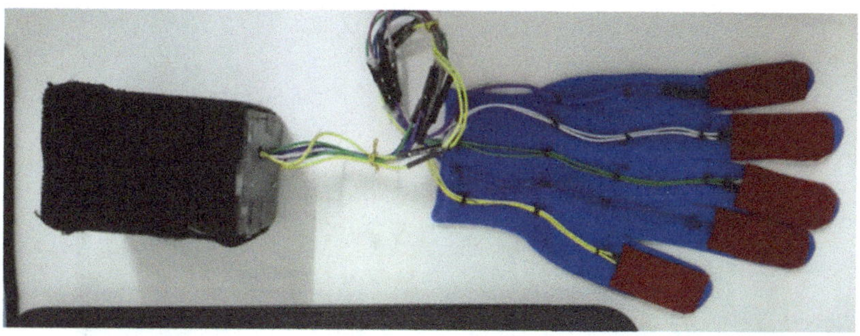

Fig. 2. Haptic glove

LMC uses infrared sensors, which requires light color to increase the accuracy of the result for the glove as it can reflect more light. In order for the user to perceive the haptic feedback, the glove need to be as fit as possible so all the sensors can be in contact with user's hand and fingers, hence, nylon glove was used in this project.

3.2 Designing the Virtual Environment in Unity3D

Unity3D software is one of the game engines that used to create a virtual environment. The software provides a powerful tool for creating a high realistic computer graphics mainly for game applications and have the features of fast rendering and real-time interaction. LMC, which is also used in the experiment allows detection and tracking of the hand's movement in real-time. Unity3D comes with a SDK library that provides the hand animations which is often required for VR environment. Using this SDK, virtual hands are displayed in the VE and is designed to resemble the human's hand.

The environment is a virtual 3D world consisting of a virtual human hand and five cubes of different sizes and colors. Unity3D also contains features such as collision to that allows interactions between objects in VE which simulates real-world situation. As the LMC have a limited distance area to track hand's movement, the virtual object also required to be designed within the cycle of the virtual hands can reach inside Unity. The game object (cube) is equipped with rigid body with following settings; gravity is turns on with the mass set to one while the rest is set to default. Table 1 shows the size of cube according to the color (Fig. 3).

Table 1. Description of the cubes

	Cube 1	Cube 2	Cube 3	Cube 4	Cube 5
Color	Blue	Orange	Yellow	Red	Purple
Size	$1 \times 1 \times 1$	$1.5 \times 1.5 \times 0.5$	$1.5 \times 1.5 \times 1.5$	$4 \times 3 \times 1$	$5 \times 5 \times 5$

Fig. 3. Virtual environment and virtual hands set up

4 Experiments

There are two parts of experiments involved in this paper. The first part consist of a preliminary observation to test the haptic glove. We implement a vibration to determine collision between the glove and the objects in the environment and therefore in this experiment, we tested the rate of this vibration. Each cube was tested for every finger on the right hand; thumb, index, middle, ring and pinky. Each test were repeated five times and the results are averaged.

After a satisfactory observation, we investigate the user testing and acceptance of haptic device which is done in the second part of the experiment. The main objective is to evaluate the effectiveness of the haptic glove in VE in enriching user's experience. The evaluation was conducted by observing the users based on two parameters: (i) does the haptic feedback improve the performance and the realism of the VE; and (ii) is a sense of presence can be achieved by wearing the haptic glove.

Before conducting the experiment, we want the participants to be familiar with the device so we allow them to explore the haptic glove device and the leap motion controller. Questionnaires were also given in order to determine their knowledge on the devices and components used. Due to the limit of distance for hand-tracking by the leap motion controller, every participants is only allowed to stay in one place.

Two tasks were conducted in the second experiment; one without the haptic feedback system and other with the haptic feedback system. This allows the user to differentiate two different experiences in every task.

At the start of the gameplay, virtual hands were displayed on the virtual environment scenes to mimicked the user's hand pose. In the first experiment, where the task is to be done without the use of haptic glove, the virtual objects contains no haptic feedback system and therefore the focus is entirely on the view from the screen. As for the second experiment, users are connected to the haptic devices with the main objective of testing the efficiency and functionalities of every sensors on the fingertips in touching the virtual cubes. The haptic feedback system shall be triggered whenever the hand came in contact with the virtual cubes. This feedback comes in the form of vibration and the presence of the objects inside the virtual environment is obtained. After the testing session is done, questionnaires were distributed again to obtain their feedback of their experiences.

5 Results and Discussion

The data obtained is recorded and observed. The first part discusses the observed vibration rates of the haptic device and the second part discusses the user testing and acceptance to haptic device.

5.1 Observation on Vibration Rates of Haptic Device

During the experiment, we noticed that there were time limits to touch the virtual cube before it disappears. These time limits vary according to their sizes: the smallest cube, cube 1, with the size $1 \times 1 \times 1$ have the minimum time limit of 9.57 s while the largest cube, cube 5 with size of $5 \times 5 \times 5$ have the longest time limit of 39.69 s. Cube 2, cube 3 and cube 4 have the time limit of 11.28 s, 24.62 s, and 29.83 s respectively. Figure 4 shows the time rate for all the cubes. Based on this graph, we observed that the time rate increases as the size of the cube is increases.

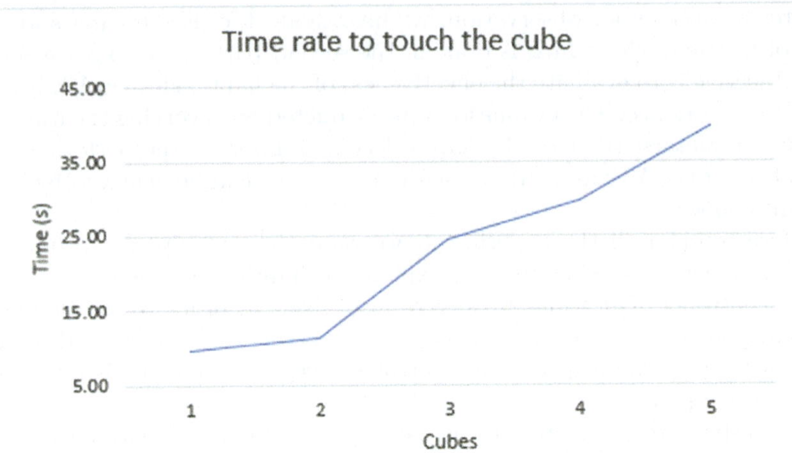

Fig. 4. Time rate to touch the cube

Within these time limit, we managed to record the vibration rate from each of the fingers for touching each cubes. From our observation, the vibration rates of touching cube 1 are 1.60 Hz/s for both thumb and index fingers, 1.00 Hz/s for middle finger, 0.50 Hz/s for ring finger and a very low value of 0.00 Hz/s for pinky. The reason for the nil value is probably because the cube is a little bit too small for the hand that it does not touch the pinky.

If we observe the vibration rates for cube 5 with 9.60 Hz/s for the thumb, 11.00 Hz/s for index finger, 10.00 Hz/s for middle finger and 8.00 Hz/s for both ring finger and pinky. The value is quite big compared to cube 1. A reason may be because the cube is large enough so it is easier and convenient to touch. Another reason is that this is probably how the SDK for Leap Motion indicated the size of virtual objects. Due to the default design of the virtual hands from the SDK, to navigate the dimensions can be challenging as it works best when the haptic glove is parallel to the leap motion controller. Less precision of the virtual hands to mimic the real hand may also be the reason for the result obtained. The rest of the vibration rates are tabulated in Table 2.

As illustrated in Fig. 5, the vibration rate increases as the size of the cube increases. The graph also shows that, index, middle and thumb fingers often have more contact with the virtual objects.

5.2 User Testing and Acceptance on Haptic Device

A group of seven undergraduate students volunteered to participate in the experiment, of which five of them were female and the other two are males. Among them, three have background study of Technology and Computing, two from Engineering and the rest are from other fields.

The results shows that the presence of haptic device in the second experiment increase the interactivity and acknowledge the user with the presence of the

Table 2. Vibration rate per cube

	Thumb	Index	Middle	Ring	Pinky
Cube 1	1.60	1.60	1.00	0.50	0.00
Cube 2	2.80	2.40	3.20	3.40	3.40
Cube 3	3.30	3.90	6.10	4.45	3.70
Cube 4	7.00	8.80	7.80	6.00	5.00
Cube 5	9.60	11.00	10.00	8.00	8.00

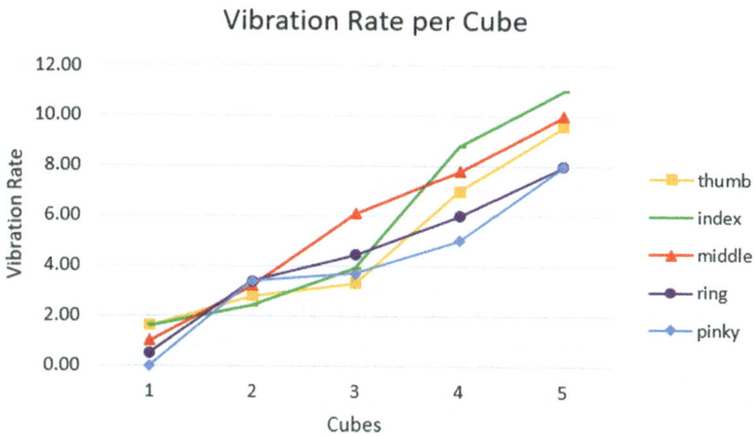

Fig. 5. Vibration rate per cube

virtual object. With haptic glove, it gives better interaction with the virtual environment compared to the first experiment and user able to feel the sensation of touch through vibration whenever cube was touched.

Interactivity, realism, enhance learning experience, ease of use, touchless interaction, motivation and fun in learning the Haptic feedback in VE were measured using questionnaires designed based on Like scale range from 1 to 5, 1 being strongly disagree and 5 strongly agree. The feedback collected from students was tabulated in Table 3.

Although it was a very small sample, it is evident and very much in line to the fact that experiential in haptic feedback system in virtual environment can have a positive impact to enhance learning experience.

Table 3. Feedback from students using the haptic glove prototype

	Description	Strongly disagree to disagree (%)	Neutral (%)	Agree to strongly disagree(%)
1	Increase interactivity			100
2	Increase realism		29	71
3	Preference on haptic devices		29	71
4	Ease of use	43	57	
5	Touchless interaction	14	29	57
6	Motivate usage of VE			100
7	Fun to use			100

6 Conclusions and Future Work

The haptic technology involves simulating the cutaneous input and output. The sensations of the skin on the fingers are obtained in the form the vibrotactile feedback. The vibration rates of the haptic device were measured in the first experiment. The information of vibration feedback was recorded; we noticed a positive correlation between the size of the virtual cube and the rate of vibration feedback on the haptic device. There was also a positive correlation between the time taken to touch the cube and the size of the cube. This side effect might be exploited as haptic perception of weight. However, more study is needed.

For the second experiment, two comparison tasks were carried out. The tasks were carried out with the haptic feedback switched off in one setup and on in the other setup. The result shows the haptic glove could improve the interactivity and the realism between user and VE. However, as the devices were connected with one another by wire, it limits the movement of the users. The leap motion controller has difficulties in detecting the haptic glove at certain period due to the fabric used.

In the future work, we will further improve on user experience by improving touch precision as well as enriching different activities such as grab, hold, translate, scale up and down, and twist the objects. The virtual environment can also be enriched by including objects having a variety of physical properties, e.g., shapes, sizes, weights and frictions.

Acknowledgments. We would like to thank Dr. Phon-Amnuaisuk for his support and guidance through the course of our project. We would also like to express our gratitude to Universiti Teknologi Brunei for the opportunity to pursue our studies here.

References

1. Pujol, L.: Realism in virtual reality applications for cultural heritage. Int. J. Virtual Real. **10**, 41 (2011)

2. Nash, E.B., Edwards, G.W., Thompson, J.A., Barfield, W.: A review of presence and performance in virtual environments. Int. J. Hum. Comput. Interact. **12**(1), 1–41 (2000)
3. Perret, J., Vander Poorten, E.: Touching virtual reality: a review of haptic gloves. In: ACTUATOR 2018; 16th International Conference on New Actuators, pp. 1–5 (2018)
4. Saddik, E.: The potential of haptics technologies. IEEE Instrum. Meas. Mag. **10**(1), 10–17 (2007)
5. Reed, C.L., Ziat, M.: Haptic perception: from the skin to the brain 73. In: Reference Module in Neuroscience and Biobehavioral Psychology. Elsevier (2018)
6. Ariza, V.Z.P., Santís-Chaves, M.: Haptic interfaces: kinesthetic vs. tactile systems (2017)
7. Silva, D.: DIY haptic glove for VR. Accessed 20 Mar 2020
8. Scheggi, S., Meli, L., Pacchierotti, C., Prattichizzo, D.: Touch the virtual reality. In: ACM SIGGRAPH 2015 Posters on - SIGGRAPH 2015. ACM Press (2015)
9. Gupta, A., O'Malley, M.K.: Design of a haptic arm exoskeleton for training and rehabilitation. In: Dynamic Systems and Control, Parts A and B, ASMEDC, January 2004
10. CyberGlove System: Cyberglove system (1990). Accessed. 15 Mar 2020
11. VrGluv: Vrgluv (2017). Accessed 15 Mar 2020

Graph Analytics to Reason Citations of Prophets in the Holy Quran

Ibrahim Venkat[1(✉)], H. N. Zainah Siau[1], M. Saiful Omar[1], Shaik Abdullah[2], and Saad Alharbi[3]

[1] School of Computing and Informatics, Universiti Teknologi Brunei, Gadong, Bandar Seri Begawan, Brunei
{ibrahim.venkat,zainah.siau,saiful.omar}@utb.edu.bn

[2] School of Distance Education, Universiti Sains Malaysia, Gelugor, Pulau Pinang, Malaysia
abdullah@usm.my

[3] Computer Science Department, Taibah University, Medina, Saudi Arabia
salharbi20@gmail.com

Abstract. Graph Analytics is a recently emerging field that offers sophisticated mechanisms to model, analyse and visualise data in the form of graphs. Citations and valuable information about various prophets have been quoted in several chapters of the Holy Quran in textual form. In this contribution, we aim to model these citation data and information using a novel graph analytics oriented framework. This paper briefly reports our findings on formulating the proposed model, data collection and empirical analysis.

Keywords: Holy Quran · Prophets · Graph theory · Data analytics · Visualisation

1 Introduction

The theoretical foundations of *Graph analytics* has been laid by *Graph theory* which gained historical importance during the eighteenth century when Leonhard Euler reasoned the Konigsberg bridge problem using the notion of a Graph. Figure 1 shows the visual representation of the problem and its associated graph formulated by Euler [3,16]. Euler represented the real world scenario of the problem: the two islands A and B; the two river banks C and D of the river Pragel; the seven bridges that provide access to A, B, C and D using a connected graph (see Fig. 1). This enabled him to express the problem statement in the context of a graph and reason the feasibility of the solution as follows: If a person X starts at any vertex and passes through each edge once (without repetition) and traverses the whole graph, is it possible for X to reach the starting vertex? Euler scientifically proved that such a solution never exists and presented his idea in a paper, which is considered to be the birth of graph theory alternatively known as graph algorithms. This has motivated scientific communities from diverse domains to

W. S. H. Suhaili et al. (Eds.): CIIS 2020, AISC 1321, pp. 192–201, 2021.
https://doi.org/10.1007/978-3-030-68133-3_19

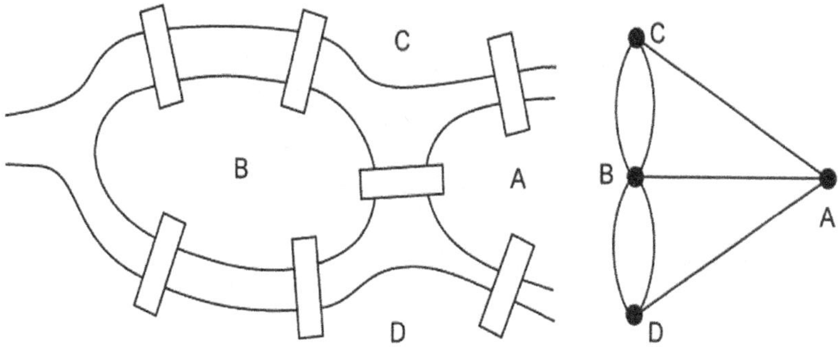

Fig. 1. Euler's representation of the Konigsberg bridge problem (real world scenario) using a connected graph

model and reason several potential problems in the context of graphs. Graph analytics which basically bridges ideas from graph algorithms and data analytics has found potential applications in several fields including E-commerce [13], biometrics [10,11,19–21], health informatics [7], distributed GPU frameworks [8], high performance computer architecture [17], citation analysis [18,22,23] and so on.

It is believed that Holy Quran, a Divine inspiration, was revealed to Prophet Muhammad through the archangel Gabriel during the period 607 CE to 632 CE [6,15]. Holy Quran is comprised of 114 chapters of varying length with two main categories namely the Meccan (the first 86 chapters) and the Medinan (the last 28 chapters). To make the reading easier, the Holy Quran has been further divided into 30 distinct sections called Juz (Para). The names of the 25 prophets that have been cited in the Holy Quran in several places is shown in Fig. 2 in a hierarchical order [1]. Holy Quran and its related literature is mostly available in textual form. The objectives of this research contribution can be summarised as follows:

i. To gather and compile the citation data in the form of a dataset (ASCII format) suitable for empirical analysis.
ii. To model the citation data of 25 Prophets with reference to the Holy Quran in the context of a *Graph Analytics Framework*.
iii. To design and develop a research prototype with GUI features that can access and explore the dataset based on user preferences and automatically generate useful graphs and analytics report.
iv. To highlight typical open problems associated with this research.

The rest of the paper is organized into the following sections: Sect. 2 briefly discusses related work pertaining to Qur'anic research; Sect. 3 presents the formulation of the proposed model; Sect. 4 provides information about the data collection, experimental setup and discussion of results; and finally Sect. 5 concludes the paper with highlights on possible open problems meant for future research.

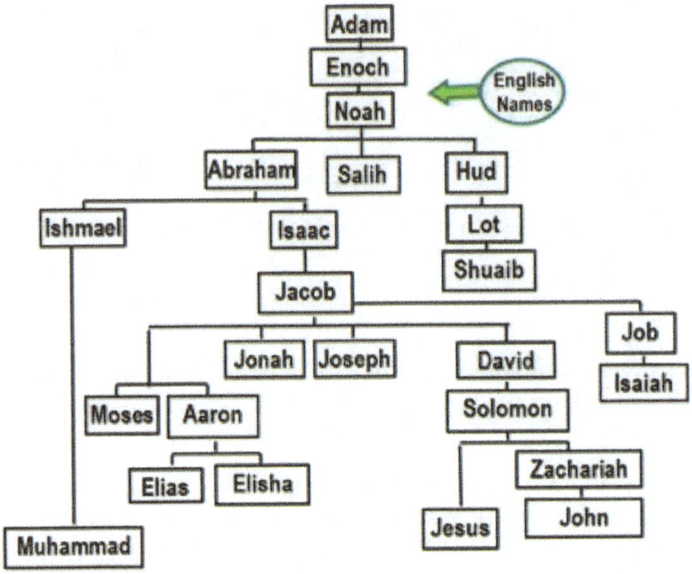

Fig. 2. Hierarchy diagram of 25 Prophets mentioned in the Holy Qur'an, courtesy of *Islamic studies resources* [1]

2 Related Work

This section intends to provide a brief overview of the investigation of computational approaches pertaining to Quranic research. Alhawarat et al. [4] have applied text mining techniques to process text data associated with the Holy Quran. Initially, preprocessing of Arabic text data has been performed to normalize the variations imposed by the prefixes and suffixes of the Arabic language. Then the normalized corpus data has been subjected to appropriate frequency transformations to generate useful statistical information. Certain statistical characteristics such as Quran chapters with high term-frequencies and pictorial visualization of the most frequently used words via intuitive wordcloud themes have been presented as a result of relevant empirical analysis. Experiments carried out in this research is based on term matrices which are built using two types of partitioning methods viz., chapters and parts. It has been observed that the choice of the partitioning methods could affect the distribution of the terms of the Holy Quran and their associated frequencies. The authors have used the Quranic data repositories offered by the Tanzil project [2] and used the R language to carry out their experimental simulations.

Noordin et al. [12] have proposed *Knowledge Representation* (KR) models pertaining to the English version of the Holy Quran. The authors have described two construction processes to build the KR models viz., ontology-based and graph database-based. Quranic Arabic Corpus has been used as the primary source to build both the KR models. It has been demonstrated that the models can serve as an information resource that can be deployed by information retrieval applications. The ontology based KR has been implemented using the *Protege* open-source tool. To construct the graph-based KR model the authors have used the *Neo4J* graph-database in conjunction with *Java*. The authors are of the opinion that the performance of both the models could be evaluated in future using an appropriate question answering system.

The problem of automating the evaluation of Holy Quran recitations has been approached using a *Pattern Recognition* (PR) framework by Rashid et al. [14]. The authors have proposed a PR-based technique to classify audio data pertaining to Quran recitations from respondents into appropriate evaluation metrics (rating scales). In the context of Turing test, basically this contribution has attempted to address a potential research problem: "Can machines (computers) evaluate Holy Quran recitations on a par with a Quran Teacher?". After pre-processing the audio signals, Mel-Frequency Cepstral Coefficients (MFCC) features have been extracted and fed into a Hidden Markov Model (HMM). Ultimately a log-likelihood ratio test is applied to classify the audio samples. The authors have used *Audocity*, an open source audio recording tool to gather audio samples. It has been reported that the proposed PR-based approach can reasonably classify recitations of Quranic words and associated Arabic alphabets. However, the authors have pointed out that the problem of automating the evaluation of recitation of Quranic verses using computational approaches remains as an open challenge to AI researchers.

3 Proposed Graph Analytics Framework

A graph $G = (V, E)$ consists of a finite set of vertices V and a set of edges E where each edge e is a set of two distinct vertices. If $e = \{u, v\} \in E$, then u and v are the end vertices of e, i.e. u and v are connected by e. We write $e = uv$. For a directed graph $D = (V, A)$, the arc set A is a set of ordered pairs of distinct vertices of V. For an arc $e = (u, v)$, u is called the tail and v is called the head of e. In a diagrammatic representation of a directed graph, the vertices are represented by dots and an arc (u, v) is represented by a line segment joining u and v with an arrow pointing to v. The in-degree of a vertex v is defined to be the number of arcs of the form (u, v) and is denoted by $id(v)$. Thus $id(v)$ is the number of arcs in D having v as head. The out-degree $od(v)$ is the number of arcs in D having v as tail. Further graph theoretic terminology can be referred from the book by Chartrand et al. [5].

We have acronymed the proposed *Prophets in Quran: Graph Analytics Framework* as PiQGAF. The proposed PiQGAF can be represented as a flowchart as shown in Fig. 3. This section intends to present the conceptual model of PiQGAF and details corresponding to the empirical aspects of the model including the *GUI*, the *Prophets Citation Dataset (PCD)* and so on will be presented in the subsequent experimental section (Sect. 4). Initially, we process the PCD in order to construct an adjacency matrix. The rows of the adjacency matrix represent the indices of the Quran chapters and that of the columns represent

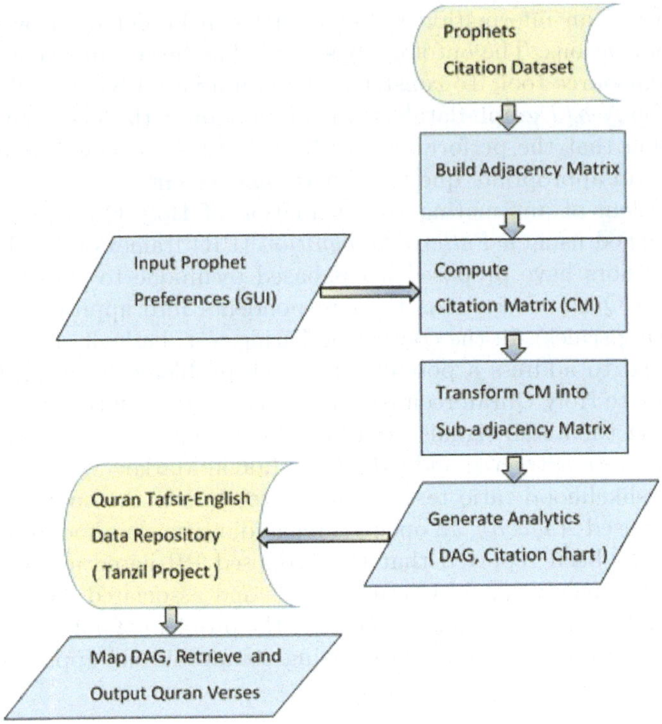

Fig. 3. Flowchart showing the sequence of processes involved in the proposed graph analytics framework

the indices of the Prophets. If a particular Quran chapter say c cites the name of a Prophet say p, the intersection of cth row and the pth column in the adjacency matrix will be set to *true*, otherwise *false*. Whenever an user makes his/her preference to chose Prophet(s) (via appropriate user interface controls), these preferences can be matched with corresponding entries in the adjacency matrix in order to extract a citation matrix. In other words, a citation matrix is basically a subset of the adjacency matrix. Hence it can be easily transformed to a sub-adjacency matrix by applying appropriate padding strategies. With the aid of this sub-adjacency matrix, elegant *Directed Acyclic Graphs* (DAGs) and citation charts can be generated for performing useful analytics and interpretations of the Holy Quran by researchers. In order to finally retrieve and output specific Quran verses which cites the chosen Prophets, the proposed PiQGAF provides sophisticated mechanisms to map the citations encoded in the DAG with the Quran Tafsir-English Data Repository [2]. In order to get some practical insight into the proposed abstract PiQGA, in the subsequent empirical analysis section (Sect. 4), we shall show example instances of user preferences, DAGs, citation charts and sample outputs.

4 Experimental Results and Discussion

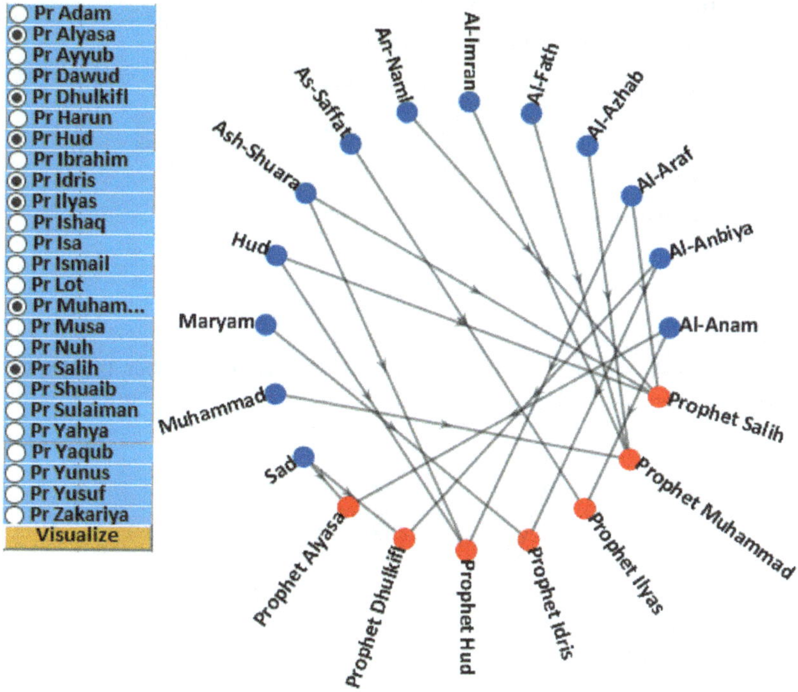

Fig. 4. An example directed graph generated by the *PiQGAF* prototype for a selection of Prophets; (Prophet Nodes) and Holy Quran *Chapter Nodes* are respectively colored as red and blue

Firstly, we have created a data set called *Prophets Citation Dataset (PCD)* as one of the contributions of this research project. Data pertaining to Quranic citations of Prophets are available in the form of tables in discrete sections of *Atlas of the Quran* compiled by the islamic scholar Shawqi Abu Khalil [9]. We have taken efforts to synthesize this bits and pieces of data into a dataset (ASCII format) suitable for computational processing. During the data collection process, it has been observed that out of the 114 chapters of the Holy Quran, citation information (mention about the names of the Prophets) is embedded in 49 chapters and the total count of citations amounts to over 500. Citation data pertaining to all the twenty-five Prophets have been organized in the format $< Prophet\ \ Name >$ followed by (newline) $< Chapter\ \ name; >$ followed by $< comma\ \ separated\ \ verse\ \ numbers >$. Chapter numbers and verse numbers are separated by a semicolon. We have implemented the abstract concepts of the proposed PiQGAF detailed in Sect. 3 in the form of a research prototype using *MATLAB* with appropriate GUI features. The directed graph generated

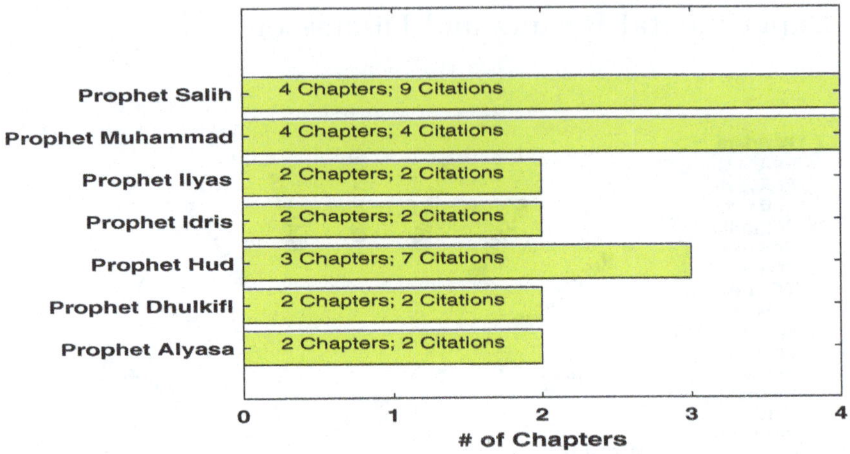

Fig. 5. An example citation chart associated with Fig. 4; Number of chapters and citation counts are automatically computed and embedded in the bars

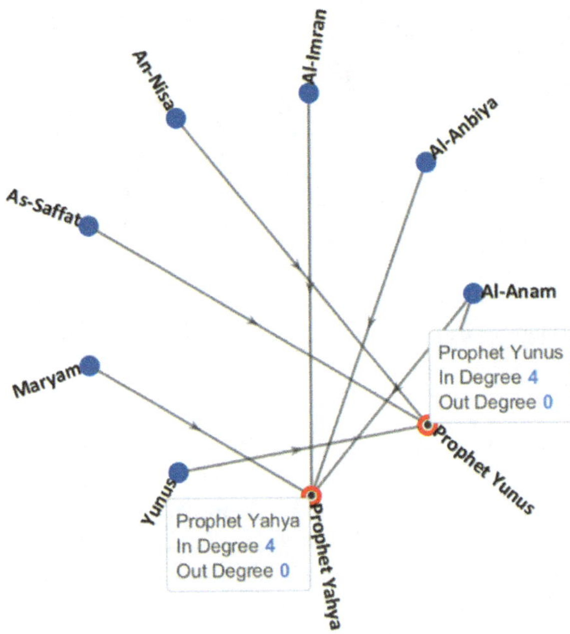

Fig. 6. Visualization of the directed graph for a typical user preference of two prophets. The *in-degree* of the prophet nodes indicates the number of citations

by the prototype for a chosen set of Prophets (user preferences via UI controls) is shown in Fig. 4. The red and blue nodes represent the names of the chosen Prophets and the corresponding chapters that cites those Prophets respec-

Prophet Yunus
4(163) We have sent thee inspiration, as We sent it to Noah and the Messengers after him:
we sent inspiration to Abraham, Isma'il, Isaac, Jacob and the Tribes, to Jesus, Job, Jonah,
Aaron, and solomon, and to David We gave the Psalms.
6(86) And Isma'il and Elisha, and Jonas, and Lot: and to all We gave favour above the nations:
10(98) Why was there not a single township (among those We warned), which believed,- so its faith
should have profited it,- except the people of Jonah? When they believed, We removed from
them the penalty of ignominy in the life of the present, and permitted them to enjoy
(their life) for a while.
37(139) So also was Jonah among those sent (by Us).

Prophet Yahya
3(39) While he was standing in prayer in the chamber, the angels called unto him:
"Allah doth give thee glad tidings of Yahya, witnessing the truth of a Word from Allah, and
(be besides) noble, chaste, and a prophet,- of the (goodly) company of the righteous."
6(85) And Zakariya and John, and Jesus and Elias: all in the ranks of the righteous:
19(7) (His prayer was answered): "O Zakariya! We give thee good news of a son: His name shall be
Yahya: on none by that name have We conferred distinction before."
19(12) (To his son came the command): "O Yahya! take hold of the Book with might": and
We gave him Wisdom even as a youth,
21(90) So We listened to him: and We granted him Yahya: We cured his wife's (Barrenness) for him.
These (three) were ever quick in emulation in good works, they used to call on Us with love
and reverence, and humble themselves before Us.

Fig. 7. Sample output retrieved from the Quran Tafsir-English Data Repository in conjunction with Fig. 6

tively. The prototype enables users to dynamically chose their preferred choice of prophets and visualize the citations in the form of an intuitive directed graph. The citation chart associated with the directed graph (see Fig. 4) is shown in Fig. 5. The in-degree of the Prophet nodes in the directed graph shown in Fig. 4 basically infers the number of chapters that cite those prophets. Further, the citation chart projected in Fig. 5, specifically the information (labels) embedded in the bars, displays the total number of chapters and total number of citations with respect to each chosen prophet. If certain edges in the directed graph are bit cluttered due to overlapping with other edges, the citation chart can be used as an additional reference. The directed graph and the corresponding output of the Holy Quran chapters and verses retrieved from the Quran Tafsir-English Data Repository for an example set of Prophets (Prophet Yunus and Prophet Yahya) are shown in Fig. 6 and Fig. 7 respectively.

5 Conclusion and Future Work

We perceive that modeling and implementing *Graph Analytics* techniques and tools for the problem of visualizing and reasoning the citations of Prophets in the Holy Quran is quite novel. Results and discussions presented in this paper justifies that this research project has laid the necessary fundamental theoretical and experimental foundations for the underlying problem. In particular, the Prophet Citation Dataset which contains data pertaining to 500 over citations could serve as a useful resource for future research projects. Further investigation of recent machine learning algorithms and probabilistic models to discover useful knowledge from data associated with the Holy Quran and its related domains could be an open challenge to the AI research community. Advanced graphical models such as Bayesian Networks, Markov Logic Networks could be explored for discovering knowledge from data. Another interesting open challenge could be to apply undirected graphs to model notions that involve interaction and exchange of information entailed by the nodes of the graphs. An interesting avenue for future directions could be to establish cross-disciplinary research projects that can be beneficial to the research communities of diverse domains who would like to deploy Graph Analytics for potential problems.

Acknowledgment. We would like to thank Al-Masjid an-Nabawi Library (The Prophet's Mosque, Madinah) for allowing us to access the library during our research visit. We thank the anonymous reviewers and Dr. Bakur Al Abed, Faculty of Quran, Islamic University of Madinah for going through our paper and providing valuable suggestions. We thank the Japan International Cooperation Agency (JICA) for extending their financial support.

References

1. Islamic studies resources. http://www.islamicstudiesresources.com/. Accessed 14 Aug 2020
2. Tanzil repository for Quran translations. http://tanzil.net/docs. Accessed 10 Oct 2020
3. Alexanderson, G.L.: About the cover: Euler and Königsberg's Bridges: a historical view. Bull. Am. Math. Soc. **43**(04), 567–574 (2006)
4. Alhawarat, M., Hegazi, M., Hilal, A.: Processing the text of the Holy Quran: a text mining study. Int. J. Adv. Comput. Sci. Appl. **6**(2) (2015). https://doi.org/10.14569/IJACSA.2015.060237
5. Chartrand, G., Lesniak, L., Zhang, P.: Graphs and Digraphs, 5th edn. CRC Press, Boca Raton (2011)
6. Lambert, G.: The Leaders are Coming. WestBow Press, Nashville (2013)
7. Hasan, S., Rivera, D., Wu, X.C., Durbin, E.B., Christian, J.B., Tourassi, G.: Knowledge graph-enabled cancer data analytics. IEEE J. Biomed. Health Inform. **24**(7), 1952–1967 (2020)
8. Jatala, V., Dathathri, R., Gill, G., Hoang, L., Nandivada, V.K., Pingali, K.: A study of graph analytics for massive datasets on distributed multi-GPUs. In: 2020 IEEE International Parallel and Distributed Processing Symposium (IPDPS), pp. 84–94, May 2020. ISSN 1530-2075
9. Khalil, S.A.: Atlas of the Qur'an, 1st edn. Dar-us-Salam Publications, Riyadh, New York (2003)
10. Lahasan, B.M., Venkat, I., Al-Betar, M.A., Lutfi, S.L., Wilde, P.D.: Recognizing faces prone to occlusions and common variations using optimal face subgraphs. Appl. Math. Comput. **283**, 316–332 (2016)
11. Lahasan, B.M., Venkat, I., Lutfi, S.L.: Recognition of occluded faces using an enhanced EBGM algorithm. In: 2014 International Conference on Computer and Information Sciences (ICCOINS), pp. 1–5, June 2014. https://doi.org/10.1109/ICCOINS.2014.6868371

12. Noordin, M.F., Sembok, T., Othman, R., Gusmita, R.H.: Constructing an ontology-based and graph-based knowledge representation of English Quran. Jurnal Teknologi (Sci. Eng.) **78**(8–2), 35–41 (2016)
13. Pyrkina, O.E., Zadadaev, S.A.: Application of crash-tests for graph analytics of electronic payment complex system for investigation of system economic security. In: 2019 Twelfth International Conference "Management of large-scale system development" (MLSD), pp. 1–4, October 2019
14. Rashid, N.R.R., Venkat, I., Damanhoori, F., Mustaffa, N., Husain, W., Khader, A.T.: Towards automating the evaluation of Holy Quran recitations: a pattern recognition perspective. In: 2013 Taibah University International Conference on Advances in Information Technology for the Holy Quran and Its Sciences, pp. 424–428, December 2013. https://doi.org/10.1109/NOORIC.2013.88
15. Drew, M.R., Williams, R.H.: Pendulum: How Past Generations Shape Our Present and Predict Our Future. Vanguard Press, New York (2012)
16. Arumugam, S., Ramachandran, S.: Invitation To Graph Theory. Scitech Publications (India) Pvt Limited (2006). google-Books-ID: 3OfSjwEACAAJ
17. Slota, G.M., Rajamanickam, S., Madduri, K.: High-performance graph analytics on manycore processors. In: 2015 IEEE International Parallel and Distributed Processing Symposium, pp. 17–27, May 2015. ISSN 1530-2075
18. Timilsina, M., Davis, B., Taylor, M., Hayes, C.: Towards predicting academic impact from mainstream news and weblogs: a heterogeneous graph based approach. In: 2016 IEEE/ACM International Conference on Advances in Social Networks Analysis and Mining (ASONAM), pp. 1388–1389, August 2016
19. Venkat, I., De Wilde, P.: Learning gait component relationships by fusing logic and graphs using Markov Logic Networks. In: 2010 13th International Conference on Information Fusion, pp. 1–8, July 2010
20. Venkat, I., De Wilde, P.: Robust gait recognition by learning and exploiting subgait characteristics. Int. J. Comput. Vision **91**(1), 7–23 (2011)
21. Venkat, I., Khader, A.T., Subramanian, K., De Wilde, P.: Recognizing occluded faces by exploiting psychophysically inspired similarity maps. Pattern Recogn. Lett. **34**(8), 903–911 (2013)
22. Wanjantuk, P., Keane, J.: Finding related documents via communities in the citation graph. In: IEEE International Symposium on Communications and Information Technology, ISCIT 2004, vol. 1, pp. 445–450, October 2004
23. Zafar, L., Ahmed, U., Islam, M.A.: Citation context analysis using word-graph. In: 2019 2nd International Conference on Communication, Computing and Digital systems (C-CODE), pp. 120–125, March 2019

Network Security and IoT

Anatomy of Ransomware: Attack Stages, Patterns and Handling Techniques

P. Ravi Kumar(✉) and Hj Rudy Erwan Bin Hj Ramlie

Univesiti Teknologi Brunei, Bandar Seri Begawan, Brunei, Brunei Darussalam
{ravi.patchmuthu,rudy.ramlie}@utb.edu.bn

Abstract. In the recent years, ransomware becomes one of the quickest, safest and profitable cybercrimes. Ransomware is a kind of rancorous malware which uses different techniques including cryptography to encrypt user files and then demand the user to pay ransom to decrypt back. It is a kind of cyber-attack which targets individuals, companies and even government sectors. This research paper provides insight into ransomware types, attack stages, attack vectors and handling techniques. This research paper is targeted to create awareness in ransomware for public and also for researchers to further explore.

Keywords: Ransomware types · Ransomware attack stages · Ransomware attack vectors · Ransomware prevention · Ransomware detection

1 Introduction

Ransomware is formed by combining two words "Ransom" and "Ware". Ransom is referring to payment and the ware is referring to the family of "Malicious Software" called Malware [1]. Ransomware is a kind of malevolent malware which can encrypt user files and/or lock the user out of their system and demand a ransom to decrypt the file or unlock the system. Apart from that, ransomware attack can cause also system downtime, loss of data, intellectual property theft, bad reputation for companies and data breach. As more and more businesses are moving into the internet (latest addition is education due to COVID-19 pandemic), digital assets keep increasing exponentially and that become the target for cyber attackers especially ransomware operators [2].

The first ransomware called AIDS Trojan or PC Cyborg, developed by biologist, Joseph L. Popp from USA in 1989 [3]. It was a simple trojan developed using symmetric cryptography and easy to decrypt but it sets the path for the ransomware attack. According to Kharraz, A et al. [4] and Simoiu, C et al. [5], ransomware is not very active until mid-2000. Only from 2005, ransomware started gearing up with the advancement of technologies and introduction of Bitcoin in 2009. According to [3], only from 2013, ransomware went pro using different attack vectors and more sophisticated methods. 2015 became a nightmare for users, and golden year for ransomware criminals due to the start of ransomware-as-a-service (RaaS) [3, 6]. RaaS is one of the main reasons for the increase of ransomware attacks in the last 4 to 5 years [7]. RaaS has made the ransomware process so easy for the attackers that anyone with little or no knowledge of

W. S. H. Suhaili et al. (Eds.): CIIS 2020, AISC 1321, pp. 205–214, 2021.
https://doi.org/10.1007/978-3-030-68133-3_20

ransomware can stage a ransomware attack using easy-to-deploy and scalable malware kits on a rental basis [2, 8]. According to [8], anonymity is very important in RaaS starting from selecting the infectious agent called as peddler, RaaS operator, attack process and until the ransom is settled. Due to this anonymity, Bitcoin (BTC) and The Onion Routers (TOR) are widely used by ransomware [9, 10]. Most of the ransomwares are targeted Windows based machines and recently Apple, Android and Linux servers are also targeted [10].

According to Steve Morgan, chief editor of Cybersecurity ventures, cybercrime damages will cost the world annually by 6 trillion dollars in 2021 [11]. He also mentioned, the ransomware damage contribution will be 20 billion dollars by 2021 [11]. According to the Cybersecurity ventures prediction that every 14 seconds, there will be a ransomware attack in the world in 2019 and in 2021, the attack will happen for every 11 seconds [11]. According to Lancaster, K., 85% of the ransomware attacks target Windows based system [12]. There is a detailed report on Coveware on average ransom amount, data recovery rate and ransomware vectors [13]. This report also says that the ransomware attack cost for a business entity is not just the ransom paid, but also the cost of the remediation of resources, lost revenue and the brand damage [13]. Most of the ransomware payments are done using cryptocurrencies due to anonymity. According to the Coveware report, Bitcoin is the most dominant cryptocurrency (99%) used for ransomware payment [13]. A joint market research by Cisco and Cybersecurity ventures identified healthcare, manufacturing, financial services, government and transportation are the 5 top most cyber-attacked industries over the last five years [14]. This report also predicted that media, entertainment, legal and education sector will join in the top 10 list by 2022. Rivero of Quartz ranked the biggest data breaches of all time in [15]. Yahoo at the top of the list with 3.5 billion accounts hacked followed by Marriott with 500 million accounts.

2 Types of Ransomware

There are basically four types of ransomware namely, Crypto ransomware, Locker ransomware, Leakware and Scareware [1, 7, 10, 16, 17]. Crypto ransomware attacks the victim machine through different attack vectors and encrypts valuable files and makes them not to accessible to the user and demands ransom to decrypt the file. Locker ransomware locks the user devices and make them not to accessible to them and demands ransom to unlock the devices. It doesn't encrypt any files. Leakware (Doxware) which only collect sensitive information from the victim's machine and then blackmails the victim for ransom [17] and Scareware is a lighter form of ransomware which informs the users that their files are encrypted or locked and demand ransom which actually did not do anything and just using the scare tactics [5]. The following Fig. 1 shows the different types of ransomware and their families [1, 3, 18, 19]:

2.1 Crypto Ransomware

Among the ransomware types, crypto ransomware dominates the numbers, attack methods and the devastation in total. Figure 2 shows that the most famous ransomwares like

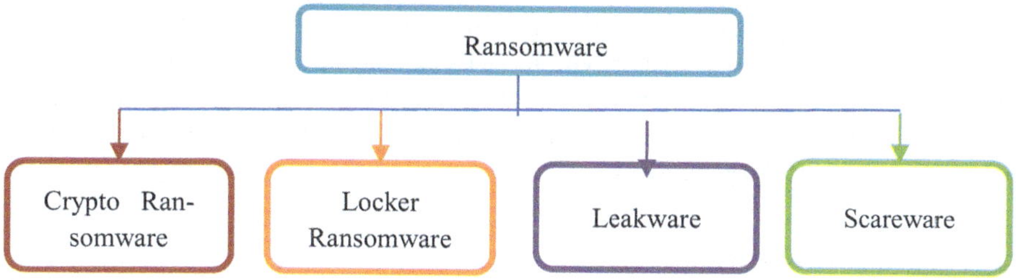

Fig. 1. Ransomware types

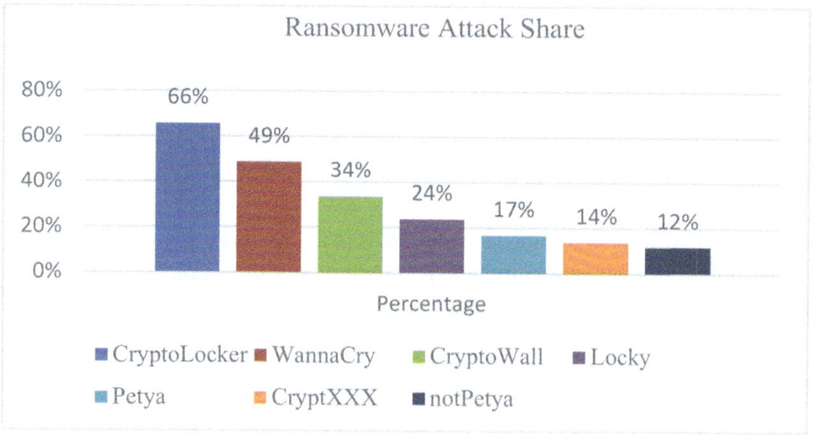

Fig. 2. Ransomware attack share

WannaCry, Locky, CryptoLocker etc., are from the crypto ransomware family and the market share [20].

WannaCry ransomware alone hit 230,000 computers across the globe in 150 countries and it caused $4 billion in losses across the globe in 2017 [21]. WannaCry ransomware targeted unpatched Microsoft Windows operating system in May 2017 using a hack called EternalBlue developed by United States National Security Agency and leaked by Shadow Brokers [21]. Microsoft released the patch two months before the attack and it affected those computers which were not patched. So, it is really important to update the patches for all the software. According to Coveware report, Sodinokibi, Ryuk, Phobos and Darma are the top four ransomwares in the Q4 of 2019 [13]. All these four ransomwares and most of the modern ransomwares are of Crypto ransomware type and this research is focused on the Crypto ransomware.

2.2 Locker Ransomware

Locker ransomwares like WinLock locks the windows screen and display porno message or fake windows product activation notice and demand ransom to unlock the window. Another one like Revton also locks the windows screen and display fake message as a law enforcement agency and demand money to unlock the window. There is also another one called Urausy Police ransomware which also lock the window and demand money to unlock it. There was one locker type in 2013 called Svpeng which locked Android devices. It makes the entire system useless and the only way to get access is to contact the attacker [22]. Here, the attacker doesn't harm or infect the computer with other malware such as password stealing etc. [22].

2.3 Leakware (Doxware)

This ransomware is a new type and it is also called as extortionware. This ransomware targets people's private valuable information like, files, photos, etc., encrypts them, holds them as hostage and demands ransom [22]. In case if the victim doesn't pay the ransom, the attackers threaten the victim that their personal data will be published online. Here, the attackers do not block access to the machine but the encrypted files are moved to a server or other infected machines [17].

2.4 Scareware

Scareware also called as fake ransomware and it is not as scary as the name indicates. It is less scary than all other ransomwares. It is technically less sophisticated and just rely on the scare tactics to coerce the victim to pay ransom [5].

3 Ransomware Attack Stages and Attack Vectors

This section explores the ransomware attack stages and attack vectors of the crypto ransomware.

3.1 Ransomware Attack Stages

After researching with many articles, ransomware white papers and online resources [1, 7, 23–27], we came out with seven stages of ransomware attack. This seven-stage ransomware attack is shown below in Fig. 3 and described briefly:

- Reconnaissance (1) – The main goal of this stage is to find the weak points of the target. In this stage, the attacker does a preliminary survey to gain information about a target such as company details, employee's details, email addresses, companies' suppliers and contractors etc. and form a dossier to use it in the next stages. Apart from that, it also collects the network, host and security policies of the target.
- Delivery and Execution (2) – In this stage, the ransomware executables are delivered to the victim's computer and start executing it. At this stage, the ransomware code ensures connectivity with its command and control (C & C) server, controlled by the attacker which is used to send commands to the compromised system. In the case of ransomware, the main purpose of communication with the C & C server is to get the encryption key. After making many changes in the system, the attacker almost owns the system and establishing persistence.

(1) (2) (3) (4) (5) (6) (7)

Fig. 3. Ransomware attack stages

- Exploitation and Infection (3) – In this stage, the attacker finalises an attack plan and infect the target machine after the preliminary survey is done. Attacker explores the different attack vectors such as spear phishing, social engineering, whaling attacks, or exploit kits and select one or more vectors and infect the target computer for further action. Here, the target computer is infected with ransomware, but the files are not encrypted. In this stage, the target computer's automated security systems like firewall, intrusion detection system, proxy and anti-virus software are deactivated by ransomware.
- Scanning and Backup Spoliation (4) – In this stage, the ransomware file scans the system for valuable files to encrypt, and also it removes the backup files and folders from the victim's machine. Here, scanning can be done on the local drives, network drives and also on the cloud (synced folders from the system).
- File Encryption (5) – Here, the ransomware will encrypt the selected valuable files using symmetric-key or asymmetric key encryption algorithms. The encryption keys will be delivered by the C & C server and encrypt the files in the local system and removes the original files.
- User Notification and Clean up (6) – In this stage, the ransomware cleans the place and wipe all the evidences of the operation and demand the user for extortion and payment. Victims are given a few days to pay the ransom and if the victim doesn't pay then the ransom will be increased and other actions will be initiated.
- Pay Process (7) – In this stage, the ransom is timed for a date that will have the most impact on the business such as on an audit day, merger day, etc. This force the victim the only option is to pay the ransom to the attackers. Ransom payments are mostly paid through Bitcoins and it is difficult to track it.

These stages are in general. It might be little different from one crypto ransomware to another. As far as the attacker is concerned, all the seven stages are important and as for as the security professionals are concerned, the first three stages, Reconnaissance, Delivery and Execution and Exploitation and Infection are important. Once the ransomware crossed the above three stages, it is very difficult to stop it. Security professionals and end users need to focus on these three stages to thwart or mitigate the ransomware attacks.

3.2 Ransomware Attack Vectors

There are different types of ransomware attack vectors. Ransomware may employ one or more attack vectors based on the attack plan to achieve its fullest objective. Each stage in the ransomware may employ different techniques to achieve its task. For example, reconnaissance uses passive reconnaissance and active reconnaissance mechanisms to collect as much information as possible about the user, the network and other activities of the targeted host [28]. After doing lots of research from journals, white papers and websites [27–31], we came out with eight ransomware attack vectors as shown below in Fig. 4. Spam/phishing email is the most popular attack vector used by ransomwares. According to [20], nearly 67% of the attack method uses spam/phishing email.

Fig. 4. Ransomware attack vectors

The Fig. 4 clearly shows eight different attack vectors. They are phishing and spear phishing email, drive-by download, malvertising, exploit kits, social engineering attacks, Remote Desktop Protocol (RDP) attacks, Bot net attacks and finally the Ransomware as a Service (RaaS). In Exploitation and Infection stage, different ransomware uses different types of exploit kits. For example, Angler is one of the popular exploit kits used in ransomwares like CryptoLocker, CryptoWall and TeslaCrypt [32].

4 Ransomware Handling Mechanisms

There are three stages in dealing with ransomware as shown in Fig. 5 They are:

- Prevent
- Detect and
- Reaction

4.1 Ransomware Prevention

Ransomware can be prevented by applying proper protection mechanisms in the early stages of ransomware attack, i.e. the first three stages, namely reconnaissance, delivery and execution, and exploitation and infection which can reduce the impact of ransomware attacks. Possibly it can even stop before the payload executes. The following are the prevention techniques we recommend after referring research articles [9, 33, 34]:

Fig. 5. Ransomware handling mechanisms

- Use strong antivirus and anti-malware software and update it regularly
- Use firewall and Intrusion Prevention System (IPS)
- Use end point protection on the network
- Never click on unverified links in the email and untrusted websites
- Never open any untrusted email attachments
- Download files only from trusted websites
- Never share your personal data to anyone
- Update patches on operating system, browser and other software regularly
- Use strong email filter and content scanner on the server itself
- Use VPN when using public Wi-Fi
- Take offline and cloud backup and always test the backup for recovery
- Don't use USB and other removable media without scanning
- End user awareness and training
- Disable all unused ports and services and avoid browser plugins and toolbars
- Give minimal access rights to users according to the roles

4.2 Ransomware Detection

Advance ransomwares can overcome the prevention methods using different techniques. The following are the techniques that can be employed by ransomware detection if it has slithered away from the prevention methods [35, 36]:

- Monitor and scan the network traffic and system for any anomalies
- Employ Intrusion Detection System (IDS)
- Employ sandboxing technique to detect dynamically new or advanced ransomware
- Employ machine learning algorithms to detect anomalies in the file system, file extensions, network traffic etc.
- Employ call-home detection method which is used to detect the communication between the infected computer and the C & C server

4.3 Ransomware Reaction

Our research recommends the following methods to recover from a ransomware attack if it has passed the previous two stages [37, 38]:

- Inform the IT Security team within the company and inform to the IT security authorities if it is individual user.

- Follow the security incident response plan
- Isolate the infected machines from the network
- Reduce the impact and the spread of the incident
- Trace the attack
- Restore the data from the clean backup and continue the business
- Don't pay the ransom

5 Conclusion

Ransomware is one of the new additions to the malware family. Ransomware is a more serious type of malware because it can cause financial damages apart from doing technical damages. With the introduction of ransomware as a service (RaaS), any unskilled individuals can involve in ransomware attack in this business. Ransomware growth is powered by Bitcoin and TOR. This makes the whole operation in the anonymous mode and makes it difficult for the security people to detect. It is becoming a multibillion-dollar industry for cybercriminals. Mission critical cloud data, Internet of Things (IoT) and critical infrastructure are the future target of ransomware. Cyber security professionals thought after the introduction of cryptojacking, ransomware will reduce but in reality, ransomware is keep increasing every year. In the future, emerging detection techniques based on artificial intelligence and machine learning can help the security professionals to handle ransomware.

References

1. Humayun, M., Jhanjhi, N.Z., Alsayat, A., Ponnusamy, V.: Internet of things and ransomware: evolution, mitigation and prevention. Egypt. Inf. J. (2020). https://doi.org/10.1016/j.eij.2020.05.003,1-13
2. Alshaikh, H., Ramadan, N., Hefny, H.A.: Ransomware detection and mitigation techniques. Int. J. Comput. Appl. **177**(40), 31–39 (2020)
3. KnowBe4 Homepage, Ransomware. https://www.knowbe4.com/ransomware, Accessed 19 July 2020
4. Kharraz, A., Robertson, W., Balzarotti, D., Bilge, L., Kirda, E.: Cutting the gordian knot: a look under the hood of ransomware attacks. In: 12th International Conference on Detection of Intrusions and Malware, and Vulnerability Assessment, Milan, Italy, 9–10 July 2015, pp. 3–24. Springer, Heidelberg (2015)
5. Simoiu, C., Gates, C., Bonneau, J., Goel, S.: "I was told to buy a software or lose my computer. i ignored it", a study of ransomware. In: 15th International USENIX Conference on Usable Privacy and Security, Santa Clara, CA, USA, 11–13 August 2019, pp. 155–174 (2019)
6. Mauyra, A.K., Kumar, N., Agrawal, A., Khan, R.A.: Ransomware: evolution, target and safety measures. Int. J. Comput. Sci. Eng. **6**, 80–85 (2018)
7. Moussaileb, R., Bouget, B., Palisse, A., Le Boudher, H., Cuppens, N., Lanet, J.L.: Ransomware's early mitigation mechanisms. In: 13th International Conference on Availability, Reliability and Security, Hamburg, Germany, 27–30 August 2018, pp. 1–10 (2018)
8. Unterfingher, V.: Ransomware as a Service (RaaS) – A Contemporary Mal du sieclu? Heimdal Security. https://www.heimdalsecurity.com/blog/ransomware-as-a-service, Accessed 04 Aug 2020

9. Krunal, A.G., Kumar, D.P.V.: Survey on ransomware: a new era of cyber attack. Int. J. Comput. Appl. **168**(3), 38–41 (2017)
10. Almashhadani, A.O., Kaiiali, M., Sezer, S., O'kane, P.: A multi-classifier network-based crypto ransomware detection system: a case study of locky ransomware. IEEE Access **7**, 47053–47067 (2019). https://doi.org/10.1109/ACCESS.2019.2907485
11. Morgan, S.: Top 5 Security Facts, Figures and Predictions and Statistics For 2020 To 2021. https://cybersecurityventures.com/top-5-cybersecurity-facts-figures-predictions-and-statistics-for-2019-to-2021/, last accessed 2020/08/07
12. Lancaster, K.: 10 2020 Ransomware Statistics That You Need to See. https://www.idagent.com/blog/10-2020-ransomware-statistics-that-you-need-to-see/, Accessed 08 Aug 2020
13. Coveware: Ransomware Costs Double in Q4 as Ryuk, Sodinokibi Proliferate. https://www.coveware.com/blog/2020/1/22/ransomware-costs-double-in-q4-as-ryuk-sodinokibi-proliferate, Accessed 08 Aug 2020
14. Morgan, S.: 2019 Cybersecurity Almanac: 100 Facts, Figures, Predictions and Statistics. Cisco and Cybersecurity ventures. https://cybersecurityventures.com/cybersecurity-almanac-2019/, Accessed 11 Aug 2020
15. Rivero, N.: The biggest data breaches of all time, ranked. Quartz. https://qz.com/1480809/the-biggest-data-breaches-of-all-time-ranked/, Accessed 17 Aug 2020
16. Kaspersky: What are the different types of ransomware? https://www.kaspersky.com/resource-center/threats/ransomware-examples, Accessed 20 Aug 2020
17. Anghel, M., Racautanu, A.: A note on different types of ransomware attacks. IACR Cryptol, ePrint Arch (2019). https://eprint.iacr.org/2019/605.pdf, Accessed 20 Aug 2020
18. Wilson,V.: 10 Ransomware Examples to Stay Away From. https://spinbackup.com/blog/10-ransomware-examples-to-stay-away-from/, Accessed 21 Aug 2020
19. Moussaileb, R., Berti, C., Deboisdeffre, G., Cuppens, N., Lanet, J.L.: Watch out! doxware on the way…. In: 14th International Conference on Risks and Security of Internet and Systems, CRiSIS 2019, Hammamet, Tunisia, 29–31 October 2019 (2019)
20. Eric, C.: Ransomware Facts, Trends & Statistics for 2020. https://www.safetydetectives.com/blog/ransomware-statistics/, Accessed 22 Aug 2020
21. Kaspersky: What is WannaCry Ransomware? https://www.kaspersky.com/resource-center/threats/ransomware-wannacry, Accessed 23 Aug 2020
22. Jelen, S.: Ransomware Attacks: What They are, Common Types and How to Protect Yourself. https://securitytrails.com/blog/ransomware-attacks, Accessed 25 Aug 2020
23. Berrueta, E., Morato, D., Magana, E., Izal, M.: A survey on detection techniques for cryptographic ransomware. IEEE Access **7**, 144925–144944 (2019). https://doi.org/10.1109/ACCESS.2019.2945839
24. Rayome, A.D.: Infographic: The 5 phases of a ransomware attack. https://www.techrepublic.com/article/infographic-the-5-phases-of-a-ransomware-attack/, Accessed 26 Aug 2020
25. Metivier, B.: Anatomy of Ransomware Attack and How to Detect the Threat. https://www.tylercybersecurity.com/blog/anatomy-of-a-ransomware-attack-and-how-to-detect-the-threat, Accessed 26 Aug 2020
26. Exabeam: The Anatomy of a Ransomware Attack, Threat Research Report. https://www.exabeam.com/wp-content/uploads/2017/07/Exabeam_Ransomware_Threat_Report_Final.pdf, Accessed 26 Aug 2020
27. Darghai, T., Dehghantanha, A., Bahrami, P.N., Conti, M., Bianchi, G., Benedetto, L.: A Cyber-Kill-Chain based taxonomy of crypto-ransomware features. J. Comput. Virol. Hacking Tech. **15**, 277–305 (2019). https://link.springer.com/article/10.1007%2Fs11416-019-00338-7
28. Deloitte: Seven Stages of Cyber Kill Chain Supplementary Reading, Cyber 101. https://www2.deloitte.com/content/dam/Deloitte/sg/Documents/risk/sea-risk-cyber-101-july2017.pdf, Accessed 03 Sept 2020

29. Sejuice: Ransomware: Attack Techniques and Countermeasures. https://www.secjuice.com/attack-techniques-countermeasures-ransomware/, Accessed 04 Sept 2020
30. Rubens, P.: Understanding Ransomware Vectors Key to Preventing Attack. https://www.esecurityplanet.com/malware/prevent-ransomware-attack.html, Accessed 04 Sept 2020
31. Aurangzeb, S., Aleem, M., Iqbal, M.A., Islam, M.A.: Ransomware: a survey and trends. J. Inf. Assur. Secur. **6**(2), 48–58 (2017)
32. Trend Micro: Exploit Kit. https://www.trendmicro.com/vinfo/us/security/definition/exploit-kit, Accessed 04 Sept 2020
33. Kaspersky: Tips on How to Prevent Ransomware Attacks. https://www.kaspersky.com/resource-center/threats/how-to-prevent-ransomware, Accessed 08 Sept 2020
34. LogRhythm: The Five Steps of Ransomware Defense. https://gallery.logrhythm.com/brochures/5-steps-of-ransomware-defense-for-sled-brochure.pdf, Accessed 08 Sept 2020
35. Newtec: Malware and Attacks Explained. https://newtecservices.ie/malware-exploit-attacks-explained/, Accessed 08 Sept 2020
36. McCormac, C.: Five Stages of a Malware Attack. https://www.sophos.com/en-us/medialibrary/Gated%20Assets/white%20papers/sophos-five-stages-of-a-web-malware-attack.pdf, Accessed 09 Sept 2020
37. NIST: Ransomware and Breach. https://csrc.nist.gov/csrc/media/presentations/ransomware-and-breach/images-media/ransomware_guidance.pdf, Accessed 09 Sept 2020
38. Mercer-Myers, C., Trueman, C.: How to Respond a Ransomware Attack. https://www.cio.com/article/3284383/how-to-respond-to-a-ransomware-attack.html, Accessed 09 Sept 2020

Low-Cost Measurement Framework for IEEE 802.11ad Communication

Shigeru Kashihara(✉), Morihiko Tamai, Akio Hasegawa, and Hiroyuki Yokoyama

Adaptive Communications Research Laboratories, Advanced Telecommunications Research Institute International (ATR), 2-2-2 Hikaridai Seika-cho, Sorakugun, Kyoto 619-0288, Japan
{shigeru,morihit,ahase,hr-yokoyama}@atr.jp

Abstract. Towards Beyond 5G and 6G, IEEE 802.11ad is spreading as the primary products in Millimeter-Wave (mmWave) communication with 60 GHz, and then it will be taken as one part of Wi-Fi systems at a fast pace. However, IEEE 802.11ad is severe in managing communication quality due to the high carrier frequency, e.g., the signal strength drops off rapidly with distance and obstacles. Since communication is sensitive to various factors such as antenna direction and distance, frequent measurement at low-cost is required to deploy a stable mmWave wireless network. In the paper, we propose a fundamental measurement framework utilizing commodity consumer devices and software, not using dedicated professional instruments. The framework employs the existing utilities for obtaining wireless and IP network information, and it then can provide many parameters for analysis. The concept contributes to one of the measurement designs for deploying stable mmWave networks in the future. Besides, the evaluation results indicated that the differences in the environment, such as antenna direction, severely impact the communication performance.

Keywords: IEEE 802.11ad · mmWave · Measurement · Beyond 5G · 6G · Framework

1 Introduction

In Beyond 5G and 6G, new attractive services such as AI-empowered applications, virtual reality, and tactile internet are expected, and then the advanced high-speed communication is essential for realizing them [1, 2]. Millimeter-Wave (mmWave) communication is one of the promising communications in Beyond 5G and 6G eras. As mmWave communication can provide multi-Gbps data rates, it will also be possible to offload the user's computation and storage resources to the edge/cloud computing server for enabling such new services.

However, since mmWave communication has an extremely high directional link due to its high carrier frequency, i.e., 30 GHz to 300 GHz, the communication quality is sensitive to various factors. Besides, the communication range is relatively short. Therefore, compared with existing communication systems such as Wi-Fi, mmWave communication is severe in managing communication quality.

W. S. H. Suhaili et al. (Eds.): CIIS 2020, AISC 1321, pp. 215–224, 2021.
https://doi.org/10.1007/978-3-030-68133-3_21

At present, IEEE 802.11ad as the primary products in mmWave communication with 60 GHz will be spreading for taking one part of Wi-Fi systems at a fast pace. However, communication characteristics are very different from the existing Wi-Fi systems. Thus, to bring out most of the potential, it is vital to understand the characteristics adequately before and after deploying a network.

So far, some papers have analyzed the communication performance of IEEE 802.11ad [3–7]. They provide valuable insights into the communication performance of IEEE 802.11ad through mathematical and experimental analysis. Mathematical analysis reveals the logical characteristics, and the experimental analysis contributes to the network design based on an actual environment. In particular, as the mmWave signal is highly directional, various factors such as antenna direction and distance impact communication performance. That is, as different environments do not have quite the same results, showing the results of the various environment are useful for us. For making a valuable comparison, we need to consider a way to evaluate the communication performance.

In this paper, we propose a fundamental measurement framework utilizing commodity consumer devices, i.e., a low-cost measurement framework for IEEE 802.11ad communication, not using dedicated professional instruments. Since IEEE 802.11ad will be spreading as well as common networks such as Wi-Fi, a low-cost measurement should be required. This paper discusses a measurement framework that utilized the existing tools and then indicates the results of IEEE 802.11ad with the method. The proposed measurement framework contributes to the design and implementation of low-cost and frequent measurement.

2 Related Work

As described in Sect. 1, since mmWave communication, i.e., IEEE 802.11ad, utilizes a 60 GHz frequency band, the communication characteristics profoundly differ from the existing Wi-Fi systems with 2.4 and 5 GHz. An access point (AP) of the current Wi-Fi system can comparatively cover a wide area such as home and office, but the signal of IEEE 802.11ad receives the strong impact from various factors, including obstacles due to the high directional characteristics. Thus, since we suppose that it is not easy to estimate the communication area based on experience and intuition, a low-cost measurement method is required to make frequent measurements. Besides, the measurement method should have standard reference measurement items for making a useful comparison with a different environment.

The references [3–8] provides analytical/simulation models and empirical results for the communication performance of IEEE 802.11ad. Reference [8] proposed an analytical/simulation model for analyzing dependencies on the contention period and the number of sectors on the MAC delay and throughput. Such analytical models help comprehend the general characteristics at the early stage, i.e., R&D stage.

After then, it is necessary for understanding empirical quality based on real environments. References [3–7] extensively investigated the communication quality of IEEE 802.11ad in real environments. They provided practical evaluation based on various characteristic aspects. References [3, 6] investigated the comprehensive general performance for commodity consumer devices of IEEE 802.11ad. In [5], they focused on a commercial smartphone and then showed communication performance characteristics for coverage, mobility, and power consumption. Besides, reference [4] examined how the blockage rich environment impacts the user association's performance and reliability. Then, in [7], TCP variants' communication performance essential on the Internet was examined.

In the above experiments, *iperf3*[1], which is one of the dominant measurement tools on an IP network, is primarily employed to evaluate communication performance. The tool provides various measurement features at no cost. As introduced in the above, we need to make various evaluations for deploying an IEEE 802.11ad network. In addition to the natural characteristics of mmWave, as different commercial products do not have the same performance, frequent and simple measurement method for IEEE 802.11ad is required. The related work gave the performance, but they do not provide consideration of the measurement way. This paper then focuses on an organization and a framework for the IEEE 802.11ad measurement.

3 Low-Cost Measurement Framework

The section proposes a measurement framework at low-cost for IEEE 802.11ad communication. As IEEE 802.11ad has more vulnerable communication than the existing Wi-Fi of 2.4/5 GHz frequency, frequent measurement at low-cost is necessary for installing IEEE 802.11ad networks in various environments such as home and office. Section 3.1 first describes an objective and requirements for the measurement, and Sect. 3.2 then proposes a measurement framework.

3.1 Objective and Requirements

A viable solution for providing multi-Gbps data rates is mmWave communication, and that is excellent for deploying Beyond 5G and 6G networks. In particular, IEEE 802.11ad is expected to be promisingly spreading as well as the existing Wi-Fi networks with 2.4/5-GHz frequency. However, due to the vulnerable communication characteristics of IEEE 802.11ad, a low-cost and straightforward measurement solution is required for frequent examination to construct a stable network.

We here discuss the objective and requirements for proposing a low-cost measurement framework. Our primary purpose is to measure the communication performance of IEEE 802.11ad without expensive special equipment because the convenient and frequent measurement is necessary for installing a stable IEEE 802.11ad network due to vulnerable communication. Thus, in our consideration, the measurement framework should satisfy the following three requirements.

(1) Measurement for commodity consumer devices of IEEE 802.11ad
(2) Obtaining fundamental common measurement data
(3) Low-cost and simple operation

[1] https://iperf.fr (Accessed on September 5th, 2020).

The existing work makes a useful measurement, but it may not provide enough explanation due to page limitations and so on. So, it would be difficult for a beginner to execute the same measurement. However, the measurement of IEEE 802.11ad should be easy and convenient because the communication coverage is small and severely suffer from various factors, e.g., antenna direction. The paper proposes a measurement framework based on the above requirements.

3.2 Design and Implementation of Measurement Framework

The section explains the design and implementation of a measurement framework. We first discuss measurement scenarios that we assume and then contemplate measurement tools that satisfy necessary measurement parameters for IEEE 802.11ad communication.

Measurement Scenarios. We assume the following three scenarios for the measurement framework.

Scenario 1 (*S1*): One access point and one station (STA)
Scenario 2 (*S2*): One access point and *n* stations
Scenario 3 (*S3*): *n* access points and *m* stations (i.e., multiple networks)

The object of *S1* is to measure essential communication performance, such as the relationship between distance and throughput. Since each vendor's product may not have the same performance, the scenario helps grasp the actual performance of products employed, as the first measurement.

Next, we would want to understand the performance in an environment where there are multiple stations. *S2* measures each station to understand the performance, including the fairness and the performance at a different position.

S3 evaluates communication performance in a multiple-network environment. Multiple networks will make an overlapped area in a real environment, causing performance degradation due to radio interference. Therefore, since IEEE 802.11ad network has a small cell, low-cost and frequent measurement is essential to deploy an IEEE 802.11ad network.

Measurement Tools and Parameters. The section explains the tools and parameters that we employed. Since our purpose is a measurement without expensive special equipment, the measurement way should be low-cost. We then utilize the following free network utilities:

- *iwconfig*: showing the parameters of the wireless network interface
- *iperf3*: active measurement tool for IP networks
- *ping*: testing the reachability over the IP networks
- *tcpdump*: capturing packets on a network interface

In the wireless network, *iwconfig* command is generally used to set the wireless network interface parameters. It can also present the measurement parameters of lower levels, such as signal level and bitrate, when the command is executed. Then, we make practical use of *iwconfig* as one of the measurement tools for wireless network information.

In the evaluation of an IP network, *iperf3* is a dominant tool for IP network measurement, and it then creates various traffics and reports the performance. *ping* command can measure RTT in addition to the reachability. Also, *tcpdump* supports analyzing the performance in detail by capturing packets if necessary. Table 1 lists the utilities' parameters, and then we can collect diverse information by using these utilities.

Table 1. Parameters obtained by utilities.

Wireless network information	
iwconfig	Bit rate, link quality, signal level, noise level, Rx invalid nwid, Rx invalid crypt, Rx invalid frag, Tx excessive retries, Invalid misc, Missed beacon
IP network information	
iperf3	Bandwidth, jitter
ping	RTT
tcpdump	Packet-level analysis

Design and Implementation. The measurement framework employs the above four utilities, as depicted in the overall design of the measurement framework in Fig. 1. In the framework, since an operation PC manages nodes related to measurement employed, it connects with the target nodes via *ssh*. The framework then makes scripts for executing the four utilities to measure the parameters listed in Table 1.

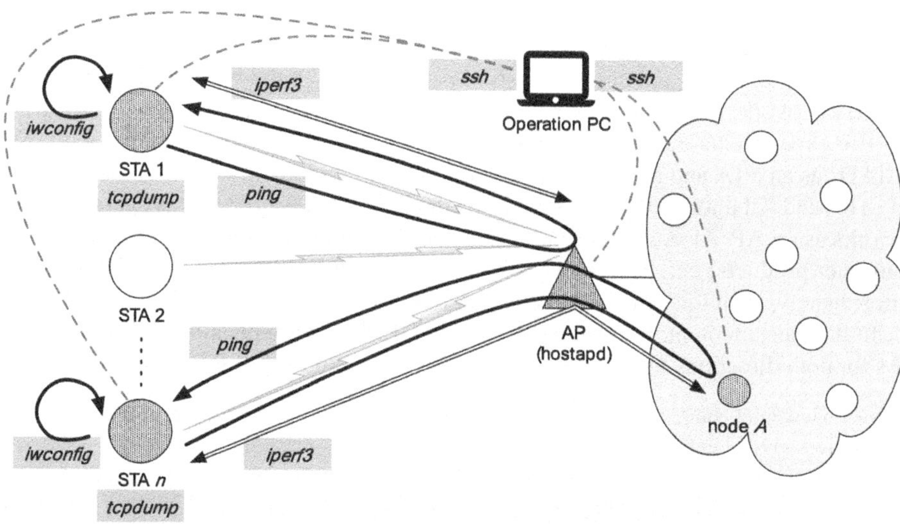

Fig. 1. Design of measurement framework

As an instance, STA1 measures the communication performance to AP. STA1 uses *iperf* and *ping* commands to AP, while *tcpdump* and *iwconfig* commands are executed for STA1. Besides, STA1 executes *iwconfig* every 100 ms to measure the wireless parameters' change, such as signal level and bit rate.

In the above explanation, if AP is a consumer product, we may not manage it. In the case, we introduce another node for measurement, i.e., node A, just behind AP, as a corresponding node to STA *n*. Besides, in the case of that STA is a smartphone or tablet, since the PC may not access it via *ssh*, we need to install the alternative application for iOS and Android.

4 Experimental Results

The section demonstrates a measurement experiment with the framework. Section 4.1 first explains experimental settings, including topology, devices, and so on. Then, Section 4.2 shows primitive results to investigate the performance characteristics. Note that since the measurement results depend on the environment, we show the instances with our measurement framework.

4.1 Experimental Settings

The paper describes the following three scenarios as instances with the measurement framework.

Evaluation 1 (*E1*): Antenna direction

Evaluation 2 (*E2*): Distance between STA and AP

Evaluation 3 (*E3*): Multiple STAs

In *E1*, since IEEE 802.11ad has an extremely high directional link, we evaluate the communication performance's impact by antenna direction. *E2* investigates the relationship between distance and communication performance as the basic performance of a wireless network. *E3* evaluates the fairness among two stations.

The experiments employ an Intel® NUC Kit NUC7i7DNHE[2] attached with SX-PCIAD[3] as an STA and an AP. Note that SX-PCIAD implements IEEE 802.11ad in the 60 GHz band. Ubuntu 16.04.6 LTS works on the PC, and then *hostapd*[4] is operated when it works as an AP. Also, although we evaluated the communication performance of the consumer products, i.e., Nighthawk® X10 Smart WiFi Router[5] and ROG Phone[6], we, unfortunately, omit the results due to page limitations.

In the measurement, we evaluate TCP and UDP's communication performance for 10 s for both directions by using *iperf3*. Besides, *ping* obtains the RTT every 100 ms for

[2] https://www.intel.com/content/www/us/en/products/boards-kits/nuc/kits/nuc7i7dnhe.html (Accessed on September 5th, 2020).

[3] https://www.silextechnology.com/unwired/60ghz-wifi-wigig-for-the-rest-of-us-sx-pcead-802.11ad-module-from-silex-brings-4.6gbps-speeds-to-new-markets-and-applications (Accessed on September 5th, 2020).

[4] https://w1.fi/hostapd/ (Accessed on September 5th, 2020).

[5] https://www.netgear.com/landings/ad7200/ (Accessed on September 5th, 2020).

[6] https://www.asus.com/Phone/ROG-Phone/ (Accessed on September 5th, 2020).

each experiment during the period, and *iwconfig* works every 100 ms. *tcpdump* captures all packets over the network interface.

4.2 Experimental Results

The section provides the above experimental results. We here present the throughput of TCP and UDP traffic. Note that the result of UDP traffic also indicates a packet loss rate.

***Evaluation 1*: Antenna Direction.** In *E1*, we investigate the impact of the antenna direction to the communication performance. In the experiments illustrated in Fig. 2, the distance between STA and AP is 50 cm, and then the antenna direction is changed for four sides (see the figure of Fig. 2). Then, Table 2 shows the communication performance for the antenna direction. In the table, Side A/A means that Side A of STA's antenna faces Side A of AP's antenna each other.

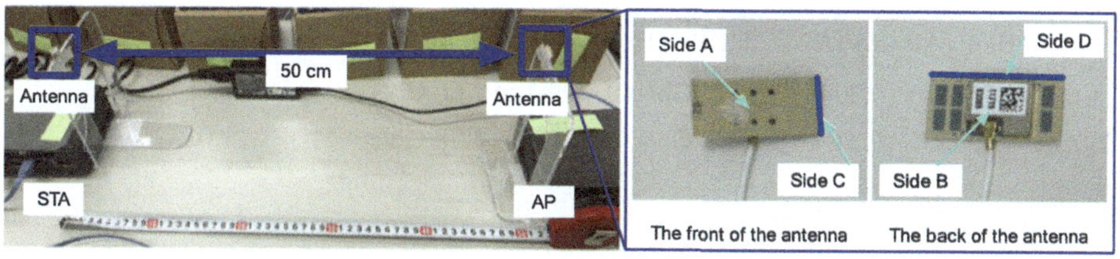

Fig. 2. Measurement for antenna direction

Table 2. Communication performance for antenna directions at the distance of 50 cm

	UDP (AP to STA)	UDP (STA to AP)	TCP (AP to STA)	TCP (STA to AP)
Side A/A	2.52 Gb/s (0.73 %)	1.95 Gb/s (4.44 %)	2.49 Gb/s	2.03 Gb/s
Side B/B	1.97 Gb/s (0.72 %)	1.96 Gb/s (1.39 %)	1.65 Gb/s	1.89 Gb/s
Side C/C	1.70 Gb/s (0.48 %)	1.70 Gb/s (0.60 %)	1.63 Gb/s	1.61 Gb/s
Side D/D	1.70 Gb/s (0.61 %)	1.73 Gb/s (1.11 %)	1.63 Gb/s	1.66 Gb/s
Side A/B	2.59 Gb/s (0.12 %)	1.77 Gb/s (1.67 %)	2.28 Gb/s	1.59 Gb/s
Side A/C	2.32 Gb/s (4.76 %)	1.67 Gb/s (1.62 %)	2.47 Gb/s	1.67 Gb/s
Side A/D	2.27 Gb/s (2.20 %)	2.07 Gb/s (5.20 %)	2.22 Gb/s	2.25 Gb/s
Side B/C	2.39 Gb/s (0.43 %)	1.74 Gb/s (0.89 %)	2.45 Gb/s	1.71 Gb/s
Side B/D	2.60 Gb/s (0.17 %)	1.89 Gb/s (6.80 %)	2.67 Gb/s	2.12 Gb/s
Side C/D	1.66 Gb/s (0.47 %)	1.66 Gb/s (0.98 %)	1.58 Gb/s	1.58 Gb/s

We can see that the antenna direction dramatically impacts on the communication performance from the results. For instance, in the experiment of TCP traffic (AP to STA), the difference between maximum and minimum throughputs is 1.09 Gb/s; that is, it is degraded by approximately 40%.

***Evaluation 2*: Distance between STA and AP.** The communication distance would be most important for deploying an IEEE 802.11ad network. Since IEEE 802.11ad uses a high frequency of 60 GHz in particular, the communication would be suddenly disconnected by the frequency characteristic as the distance increases.

Under the assumption that we investigate the communication range in our office, we evaluate the communication performance for two distances of 11 m and 23 m (Fig. 3). We also employ the antenna direction of Side A/A from the results of Table 2. Table 3 shows the communication performance of the two distances. From the results, we can see that the IEEE 802.11ad communication is available in the office. However, it can be seen that the communication performance varies greatly depending on the distance from the AP.

Table 3. Communication performance for two distances of 11 m and 23 m.

	UDP (AP to STA)	UDP (STA to AP)	TCP (AP to STA)	TCP (STA to AP)
11 m	1.75 Gb/s (0.30 %)	1.75 Gb/s (0.48 %)	1.66 Gb/s	1.66 Gb/s
23 m	1.45 Gb/s (0.37 %)	1.48 Gb/s (0.42 %)	1.28 Gb/s	1.02 Gb/s

***Evaluation 3*: Multiple Stations.** Since multiple stations connect with an AP in an existing network, the section evaluates the two stations' communication performance. In the experiment, two stations, i.e., STA1 and STA2, connect with AP, as illustrated in Fig. 4. Also, the distance among stations and AP is 2 m each other.

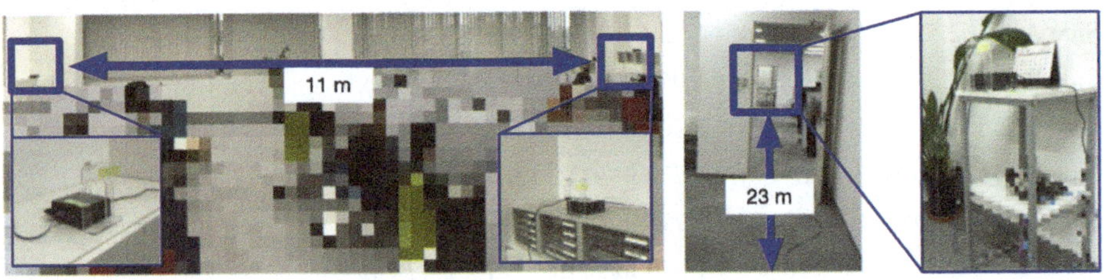

Fig. 3. Measurement for distance between station and AP

Table 4 shows the results of each station. As a result, if the distance is the same, the communication performance is almost the same. However, if the stations are located at a different distance, they will have different communication performances. Besides, when a service/application requires a specific communication quality, we need to analyze the details using the other parameters to deploy the satisfied network.

Table 4. Communication performance between two stations

	UDP (AP to STA)	UDP (STA to AP)	TCP (AP to STA)	TCP (STA to AP)
STA1	1.20 Gb/s (3.53 %)	1.08 Gb/s (4.8 %)	1.23 Gb/s	1.14 Gb/s
STA2	1.23 Gb/s (2.2 %)	1.04 Gb/s (2.08 %)	1.23 Gb/s	0.99 Gb/s

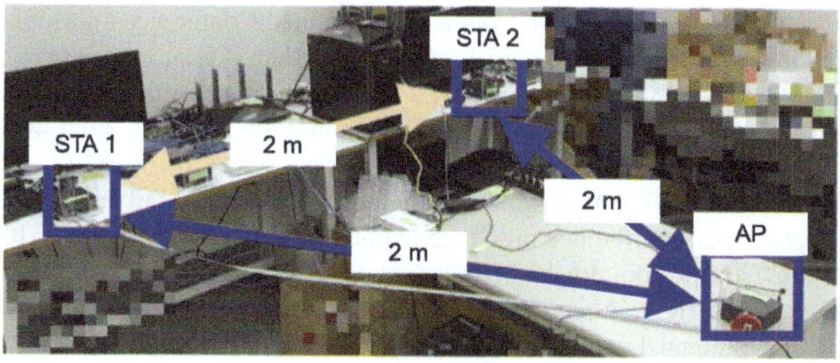

Fig. 4. Measurement for multiple stations

5 Conclusion

IEEE 802.11ad is one of the promising wireless communications in Beyond 5G and 6G eras. Compared with the existing wireless communications, e.g., IEEE 802.11 series, it provides multi-Gbps data rates. In contrast, the communication quality is easily affected by various factors such as antenna direction and distance due to the high carrier frequency and the extremely high directional link. After this, since the frequent measurement is required for the mmWave communication due to the vulnerable communication quality, the paper proposed a low-cost measurement framework for IEEE 802.11ad as one measurement framework. Besides, the evaluation results indicated that the differences in the environment, such as antenna direction, impact communication performance. In beyond 5G and 6G, an AI-enabled network optimization would be one of the challenging problems [1], and then the tremendous amount of the measurement data will be required. To realize it, we hope that the concept of the paper is one of the approaches.

Acknowledgements. This research is supported by the Ministry of Internal Affairs and Communications in Japan (JPJ000254).

References

1. Letaief, K.B., Chen, W., Shi, Y., Zhang, J., Zhang, Y.J.A.: The roadmap to 6G: AI empowered wireless networks. IEEE Commun. Mag. **57**, 84–90 (2019). https://doi.org/10.1109/MCOM.2019.1900271
2. Nawaz, S.J., Sharma, S.K., Wyne, S., Patwary, M.N., Asaduzzaman, M.: Quantum machine learning for 6G communication networks: state-of-the-art and vision for the future. IEEE Access **7**, 46317–46350 (2019). https://doi.org/10.1109/ACCESS.2019.2909490

3. Saha, S., Aggarwal, S., Assasa, H., Loch, A., Prakash, N.M., Shyamsunder, R., Steinmetzer, D., Koutsonikolas, D., Widmer, J., Hollick, M.: Performance and pitfalls of 60 GHz WLANs based on consumer-grade hardware. IEEE Trans. Mob. Comput. **1233**, 1 (2020). https://doi.org/10.1109/tmc.2020.2967386
4. Dahhani, M., Jakllari, G., Beylot, A.L.: Association and reliability in 802.11ad networks: an experimental study. In: Proceedings of the Conference on Local Computer Networks, LCN, October 2019, pp. 398–405 (2019). https://doi.org/10.1109/LCN44214.2019.8990684
5. Aggarwal, S., Thirumurugan, A., Koutsonikolas, D.: A first look at 802.11ad performance on a smartphone. In: Proceedings of the Annual International Conference on Mobile Computing and Networking, MOBICOM, pp. 13–18 (2019). https://doi.org/10.1145/3349624.3356766
6. Nguyen, K., Golam Kibria, M., Ishizu, K., Kojima, F.: Performance evaluation of IEEE 802.11ad in evolving Wi-Fi networks. Wirel. Commun. Mob. Comput. **2019**, 4089365 (2019). https://doi.org/10.1155/2019/4089365
7. Nguyen, K., Sekiya, H.: TCP behavior on multi-gigabit IEEE 802.11ad link. In: Proceedings of the 2020 International Conference on Green Human Information Technology, ICGHIT 2020, pp. 58–61 (2020). https://doi.org/10.1109/ICGHIT49656.2020.00022
8. Chandra, K., Prasad, R.V., Niemegeers, I.: Performance analysis of IEEE 802.11ad MAC protocol. IEEE Commun. Lett. **21**, 1513–1516 (2017). https://doi.org/10.1109/LCOMM.2017.2677924

Experiments on LoRa Communication Used in a Relay Station Network for Disaster Management

Kittikom Sangrit[1,2](✉), Kanokvate Tungpimolrut[1], Udom Lewlompaisarl[1], Montri Chatpoj[1], Jessada Karnjana[1], Ken T. Murata[3], Wida Susanty Haji Suhaili[4], Jennifer Dela Cruz[5], Fredmar Asarias[6], Phoummixay Siharath[7], Daoheung Bouangeune[8], and Thin Lai Lai Thein[9]

[1] NECTEC, National Science and Techlogy Development Agency, 112 Thailand Science Park, Khlong Luang, Pathum Thani 12120, Thailand
{kanokvate.tungpimolrut,udom.lewlompaisarl,montri.chatpoj, jessada.karnjana}@nectec.or.th

[2] Sirindhorn International Institute of Technology, Thammasat University, 131 Moo 5, Tiwanon Rd., Bangkadi, Muang, Pathum Thani 12000, Thailand
m6222040468@g.siit.tu.ac.th

[3] National Institute of Information and Communications Technology, 4-2-1 Nukui-Kitamachi, Koganei, Tokyo 184-8795, Japan
ken.murata@nict.go.jp

[4] School of Computing and Informatics, Universiti Teknologi Brunei, Tungku Link, Gadong BE1410, Brunei
wida.suhaili@utb.edu.bn

[5] Mapua University, Muralla St. Intramuros Manila, 1002 Manila, Republic of the Philippines

[6] Department of Science and Technology - Advanced Science and Technology Institute (DOST-ASTI), DOST-ASTI Entrance Rd, Diliman, Quezon City 1800, Metro Manila, Republic of the Philippines
fredmar.asarias@asti.dost.gov.ph

[7] Faculty of Engineering, National University of Laos, Lao-Friendship Road, Watnak Village, Sisattanak District, Vientiane Capital, Lao People's Democratic Republic

[8] Technology Computer and Electronics Institute, Sidumduan Village, Saysettha District, Vientiane, Lao People's Democratic Republic

[9] Faculty of Information Science, University of Computer Studies, Yangon, No. (4) Main Road, Shwe Pyi Thar Township, Yangon, Myanmar

Abstract. The topography of some northern provinces of Thailand is mountainous. Hence, many areas are subject to landslides. Landslides can cause severe disruption in public utilities and services, such as transportation, electricity, and telecommunication networks. Triggers for a landslide in one area often originate from remotely mountainous or rural areas far away from towns; thus, the 3G/4G system is not available. Therefore, a relay station network based on a low-power wide-area network (LPWAN) technology is proposed to monitor environmental parameters in no-3G-signal landslide-prone areas. The relay station network consists

W. S. H. Suhaili et al. (Eds.): CIIS 2020, AISC 1321, pp. 225–232, 2021.
https://doi.org/10.1007/978-3-030-68133-3_22

of an array of relay stations whose function is to forward the received data from a monitored area to the next station until the data reach the destination station. Then, the data will be sent to a server via the 3G network from the base station. To achieve the final goal, we initially conduct experiments on multi-hop communication based on the LoRa protocol. This paper aims to report and discuss the implementation of simple LoRa-based relay stations in terms of packet loss and coverage area. In our experiments, each LoRa communication module was set to minimize the communication range first so that the tests could be done within a radius of one kilometer in Thammasat University. The scope was limited by setting three parameters: transmission power, coding rate, and bandwidth. Then, a relay function was designed and developed. The performance was evaluated in terms of packet receiving rate (PRR). The experimental results showed that, for the node-to-node communication with a maximum distance of 600 m, the PRR was approximately 0.994. When a relay station was added between two nodes, the maximum range extended to 800 m, with the PRR of 0.795. These results suggest that a higher layer protocol for checking the received packets is necessary.

Keywords: Relay station network · Low-power wide-area network · LoRa-based communication · Disaster management

1 Introduction

Since 2007, the National Electronics and Computer Technology Center (NECTEC) has collaborated with the Department of Disaster Prevention and Mitigation (DDPM) of the Ministry of Interior of Thailand on a landslide monitoring project, which currently consists of weather stations and a data collection center. The weather stations, which operate on solar panels and batteries, are installed and distributed in landslide-prone, mountainous areas in many provinces, such as Chiang Mai, Mae Hong Son, Lamphun, and Lampang. The weather stations read and collect some environmental parameters, such as rainfall, temperature, and humidity. They then transmit them to the data collection center by a GSM networking connection. These data can be viewed and accessed through monitoring web pages by authorized persons. In the case of emergency (e.g., there is intense or prolonged rainfall, causing the precipitation level is higher than predefined values), the weather station sends a warning message via the short message service (SMS) to the officers who will handle the situation.

The current system relies on just the service from the GSM telecommunication, which often becomes unavailable in severe disasters such as landslides. As a consequence, the system will be down. The loss of any means of telecommunication is life-threatening when the disasters occur because the data under those devastating circumstances are vitally important for many reasons, such as aiding during rescue operations, improving mitigation strategies, or even evacuation procedures. Therefore, a reliable and resilient communication network and dedicated channel are necessary for operating in rural areas, where no mobile network is available. We name a network of nodes that transmit data from one to another as the relay station network. NECTEC and research partners including the National Institute of Information and Communications Technology (NICT), Japan, Universiti Teknologi Brunei (UTB), Brunei Darussalam, Advanced Science and Technology Institute (ASTI), Mapua University, the Philippines, National University of Laos, Technology Computer and Electronics Institute, Lao PDR, and University of Computer Studies, Yangon, Myanmar, initiated a project, titled "Relay Station Network Based on Low-power Wide-area Network (LPWAN) Technologies for Disaster Management," that aims to study, investigate, design, and implement the relay station network. The structure of the relay station network is shown in Fig. 1, and it consists of an array of relay stations that their only function is to forward the received data to the next station until the data reach the destination (base) station.

To achieve the final goal, we initially conduct preliminary experiments on multi-hop communication based on the LoRa protocol. In the proposed relay network, LoRa is chosen as the communication protocol due to its long communication range and low power consumption. This paper aims to report and discuss the implementation of a simple LoRa-based relay function regarding the received signal strength index (RSSI) and packet receiving rate (PRR).

The rest of this paper is organized as follows. Section 2 describes our implementation, providing details of the hardware we used and the communication procedures we designed. The experiments and results are presented in Sect. 3. Section 4 concludes this work.

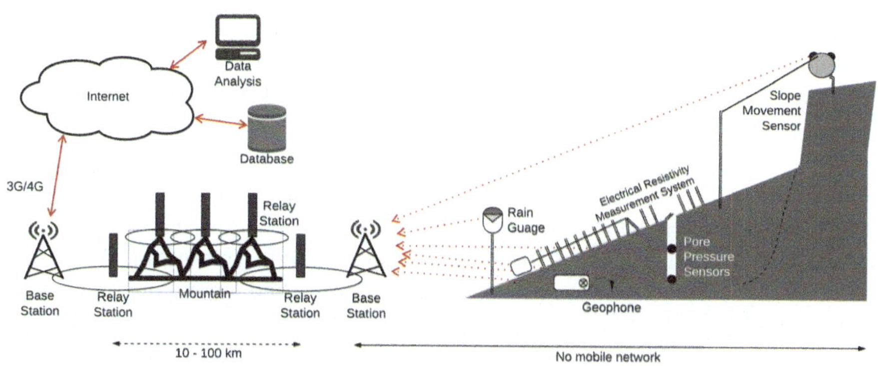

Fig. 1. Structure of the relay network (on the left) and a wireless sensor network used to collect data (on the right).

2 Implementation

For the implementation, we designed a communication procedure using a development board, called TTGO T-Beam, as shown in Fig. 2, to test the performance of the relay function. This board is manufactured by Shenzhen Xin Yuan

Electronic Technology, Co., Ltd [1]. We programmed the development board on PlatformIO, which is an alternative to the Arduino IDE. This section describes details of the development board we implemented and the communication procedure we designed.

TTGO T-Beam

The development board is equipped with the latest ESP32 chipset with a battery, providing necessary communication requirements. The board consists of WiFi and Bluetooth modules for wireless communication along with battery management, a GPS module with a ceramic antenna and a LoRa transceiver that operates at different frequencies: 433, 470, 868, 915 MHz.

Communication Procedure Design

The communication procedures are iullstrated in Fig. 3, in which the numbers indicate steps of how the transeivers interact in a sequence. Figure 3(a) shows the case of the communication between two nodes without a relay. In contrast, Fig. 3(b) shows the scenario with a relay.

Transmitter: The transmitter broadcasts a packet that includes the packet counter, GPS location, and SYN flag set for every 5 s and waits for an acknowledgment either from the receiver or relay, as shown in Fig. 3(a) and Fig. 3(b), respectively.

Receiver: The receiver records the received signal strength (RSSI) and signal-to-noise ratio (SNR) of all packets. It will send ACK if it receives a packet from a transmitter; otherwise, it will respond by sending RE-ACK, as shown in Fig. 3(a) and Fig. 3(b).

Relay: The relay extends the communication range by simply forwarding packets with the SYN flag sent from the transmitter until they reach the receiver or with the RE-ACK flag replied from the receiver back to the origin. An array of relays forwarding packets sequentially can be established to expand the communication range.

3 Experiments and Results

3.1 Communication Parameter Setup

The LoRa transceiver can be fine-tuned in many different ways by varying communication parameters to meet required conditions for a specific purpose. The physical setting parameters affect energy consumption, range and communication quality. The parameters include transmission power (TP), carrier frequency (CF), spreading factor (SF), bandwidth (BW), and coding rate (CR). With these parameters, the transceiver can be configured in more than 6720 possible combinations [2].

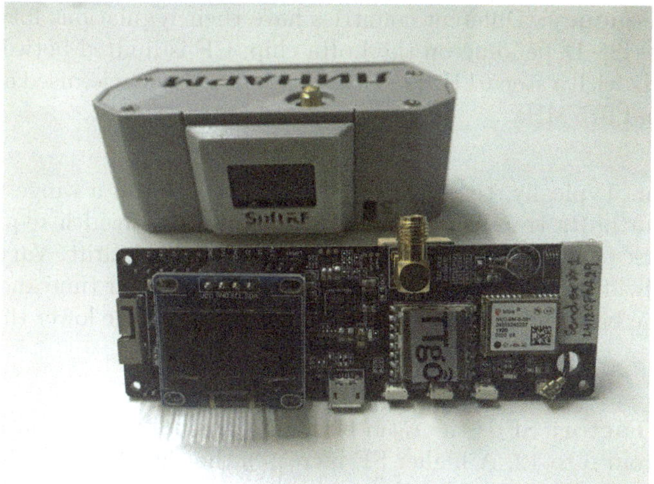

Fig. 2. TTGO T-Beam boards.

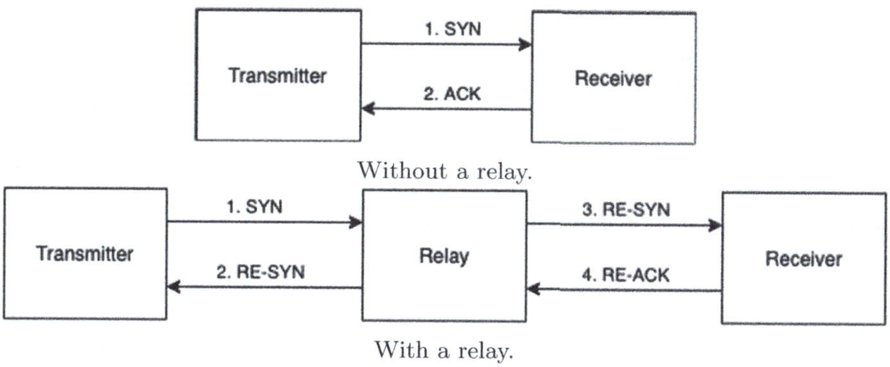

Fig. 3. Communication procedures.

Transmission Power: A typical setting of LoRa's transmission power can be adjusted, ranging from −4 to 20 dBm. However, the range varies with different hardware. Because of this hardware limitation, we can choose the range from 2 to 20 dBm [2]. Technically, increasing transmission power would increase not only communication quality and range but also energy consumption [2,3].

Coding Rate: The higher the coding rate, the greater the probability of successful reception. Unfortunately, this also increase the message length, on-air time, and energy consumption [2,3]. Note that LoRa transceivers with different coding rates can still communicate to each others [3].

Carrier Frequency: Different countries have their regulations for allowed frequency usage [2]. Depending on the LoRa chip, CF is limited between 137 MHz and 1020 MHz with a step 61 Hz [2]. However, most frequencies used are 433 MHz, 868 MHZ, and 915 MHz.

Bandwidth: Typically, the bandwidth can be chosen in a range from 125 to 500 kHz for a LoRa transceiver. Note that a lower bandwidth requires a more accurate clock to minimize problems related to the clock drift. Varying the frequency bandwidth of a LoRa transceiver affects both on-air time and sensitivity, the higher the bandwidth, the lesser the on-air time and the lower the sensitivity [2,3].

Spreading Factor: SF is the ratio between the symbol rate and chip rate. It can be set from 6 to 12. A higher SF increases the signal to noise ratio; hence, the receiver sensitivity, range, and on-air time are increased. Note that different SF values are orthogonal to each others so that network communication channels can be separated by using different SF [3].

3.2 Experiments

In our experiments, the parameters are set as follows: TP = 2 dBm, CR = 8 Hz, BW = 500 kHz, and SF = 7.

Experiment in Laboratory Environment: We tested the communication procedure we designed in our laboratory to evaluate the performance of the relay function. The experiments were conducted in three circumstances. For the first circumstance, there were only two nodes: a transmitter and a receiver communicating with each other. For the second, the distance between the transmitter and the receiver is increased. A relay is placed in the middle between two ends. For the third, we repeated the second experiment by adding one more relay so that we had two relays forwarding data from the transmitter to the receiver. We evaluated the performance by calculating the packet receiving rate (PRR) from data recorded by the receiver and the number of corrupted data, which is incompleted, received data that cannot be successfully decoded. The evaluation results are shown in Table 1. It can be seen that, as we added more relays into the communication channel, the PRR decreased slightly.

Table 1. Comparison of PRR and corrupted data in three scenarios.

No. of relays	PRR	Corrupted data
0	0.994	0
1	0.795	0
2	0.678	0

Experiment in Field Environment: We carried out an experiment to set the parameters to cover an appropriate communication area in Thammasat University and to evaluate the performance of the relay function when the network was deployed in outdoor and noisy environment. The receiver was installed at the TU Dome building, as marked by a red circle in Fig. 4, and the relay was located at a bus stop in Thammasat University, as marked by a green circle in Fig. 4. Note that there was no communication obstacle between the receiver and the relay. The transmitter was deployed and relocated at locations that were 100, 200, 300, and 400 m far away from the relay in the direction towards Thailand Science Park, where there were some obstacles between the relay and the transmitter. The experimental result showed that the communication range could be extended up to 800 m when the communication between the receiver and the relay was established with a clear line-of-sight. However, when the line-of-sight was not clear, the range was limited to approximately 600 m.

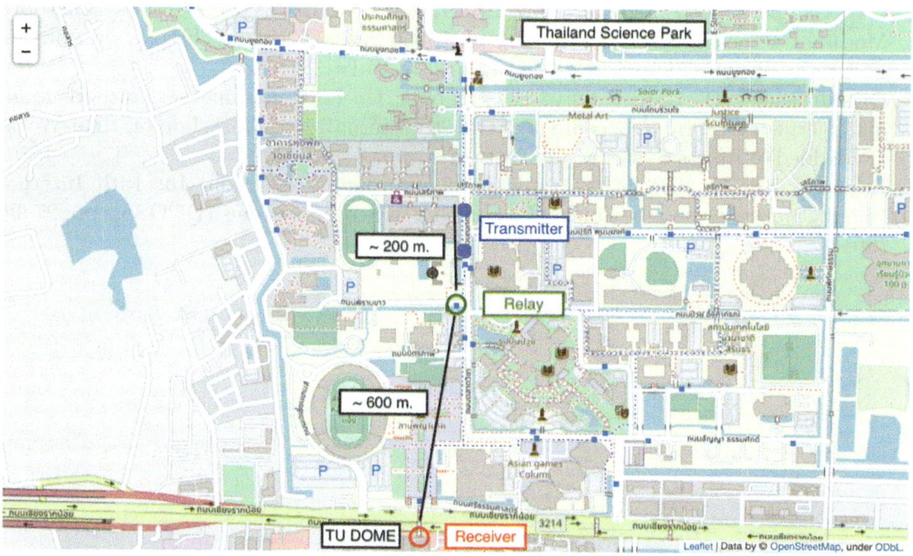

Fig. 4. Locations of the receiver, relay, and transmitter.

4 Discussion and Conclusion

This paper reported the design, implementation, and preliminarily experimental results of a network of LoRa-based relay stations, which is intended as a communication channel that transmits data from a rural area to a destination station. At the destination station, data are to be sent to a server via a 3G/4G network. This kind of relay station networks is of interest for monitoring environmental parameters in some disaster-related-prone areas. According to our experiments, the LoRa communication module was firstly set to minimize the coverage area by adjusting three parameters, which are transmission power, coding rate, and bandwidth. For the one-hop communication with a maximum distance of 600 m, the PRR was around 0.994. However, when adding a relay node to form the multiple-hop communication, the maximum range was extended with the lower PRR, which was approximately 0.795. Thus, the simple relay function worked reasonably. In order to improve the PPR further, a higher layer protocol for checking the received packets is to be developed and implemented.

Acknowledgments. This work is an output of an ASEAN IVO project, titled "Relay Station Network Based on Low-power Wide-area Network (LPWAN) Technologies for Disaster Management," and partially financially supported by National Institute of Information and Communications Technology (NICT).

References

1. LILYGO® TTGO T-Beam V0.7 ESP32 868/915Mhz WiFI Wireless Bluetooth Module GPS NEO-6M SMA LORA 32 18650 Battery Holder. http://www.lilygo.cn/prod_view.aspx?TypeId=50033&Id=1237&FId=t3:50033:3
2. Dix-Matthews, B., Cardell-Oliver, R., Hübner, C.: LoRa parameter choice for minimal energy usage. In: Proceedings of the 7th International Workshop on Real-World Embedded Wireless Systems and Networks, pp. 37–42 (2018)
3. Bor, M., Roedig, U.: LoRa transmission parameter selection. In: 13th International Conference on Distributed Computing in Sensor Systems (DCOSS), pp. 27–34 (2017)

Peatlands Monitoring in Malaysia with IoT Systems: Preliminary Experimental Results

Aduwati Sali[1(✉)], Azizi Mohd Ali[1], Borhanuddin Mohd Ali[1], Sharifah Mumtazah Syed Ahmad Abdul Rahman[1], Jiun Terng Liew[1], Nur Luqman Saleh[1], Ahmad Ainuddin Nuruddin[2], Sheriza Mohd Razali[2], Ibtisam G. Nsaif[1], Nordin Ramli[3], Hafizal Mohamad[4], and Naomi Terada[5]

[1] Wireless and Photonic Networks Research Centre of Excellence (WiPNET), Department of Computer and Communication Systems Engineering, Faculty of Engineering, Universiti Putra Malaysia, 43400 Serdang, Selangor, Malaysia
{aduwati,ma_azizi,borhan,mumtazah,liewjt,nurluqmansaleh}@upm.edu.my

[2] Institute of Tropical Forestry and Forest Products (INTROP), Universiti Putra Malaysia, 43400 Serdang, Selangor, Malaysia
{ainuddin,sheriza}@upm.edu.my

[3] Wireless Innovation Laboratory, Corporate Technology, MIMOS Berhad, Taman Teknologi Malaysia, 57000 Kuala Lumpur, Malaysia
nordin.ramli@mimos.my

[4] Faculty of Engineering and Built Environment, Universiti Sains Islam Malaysia, 71800 Nilai, Negeri Sembilan, Malaysia
hafizal@usim.edu.my

[5] National Institute of Information and Communications Technology, 4-2-1 Nukui-Kitamachi, Koganei, Tokyo 184-8795, Japan
naomi-te@nict.go.jp

Abstract. Peatland forest fires are a serious threat to the biodiversity-ecosystem and to human health in South East Asia, especially in the dry season. Here, approximately 2.13 million hectares of peat swamp forests are found and made up of 6.5% of the total land area. However, monitoring and management of sustainable peat swamp forests can be improved, as degradation of the forest may happen. One of the main challenges in the forest management process is the lack of manpower to obtain in-situ samples for analysis and monitoring purposes. This paper proposes an Internet of Things (IoT)-based peatland forest management and monitoring system in Raja Musa Forest Reserve (RMFR), Selangor, Malaysia. Real-time peatland data such as ground water level (GWL), soil temperature, and soil humidity as well as local meteorological data such as rain precipitation, wind speed, ambient temperature, and ambient humidity are obtained from deployed IoT system on sites. The peatland data from January to March 2020 are recorded and analyzed.

Keywords: Peatland forest fires · IoT system · Data analytic

W. S. H. Suhaili et al. (Eds.): CIIS 2020, AISC 1321, pp. 233–242, 2021.
https://doi.org/10.1007/978-3-030-68133-3_23

1 Introduction

Peat swamp hold as carbon pool as its biomass content is very high due to its long term accumulation and thick peat dome formation [7]. This type of forest should be continually protected as the forest is of global importance particularly due to agricultural conversion to plantations, logging, drainage and forest fires during dry seasons [14]. Serious interest in forest fire studies in ASEAN began after severe fire incidents involving peat swamp forests which occurred in 1997/1998 and caused extensive haze throughout the region. Study found large-scale fire in Indonesia released massive amount of carbon into the atmosphere which has caused serious trans-boundary haze [3].

Forest fire has now become an increasing threat responsible for burning over 370 million hectares (ha) annually on average-basis across the globe [12]. In line with this, vegetation fire statistics may be highly inaccurate. At the continental scale, Asia contributes to approximately 14% of the global total burnt area. These threatening fires are an increasingly imperative issue in South-east Asia because it has several economic, environmental and social impacts. Malaysia is no exception as statistical indication has revealed that in the last three decades, average annual temperature increased by 0.11 °C per decade together with varying in precipitation pattern within the country [4].

After several fire events, Southeast Asia Fire Danger Rating System Project was developed [4] based on the Canadian Forest Fire Danger Rating System. The objective of the system was designed to assist countries in the region in preparing earlier mitigation plan and advanced warning systems. These systems affect land and fire managers so it has long been used as a tool to address a wide range of fire management problems. This system can be described as the means of quantifying the potential or ability of a fire to start, spread, and cause damage. Formal Fire Danger Rating Systems (FDRS) have been developed in Canada, Australia, and the United States for about 75 years. The Canadian Forest Fire Danger Rating System consists of two subsystems, namely the Canadian Forest Fire Weather Index (FWI) System and the Canadian Forest Fire Behavior Prediction (FBP) System (Forestry Canada Fire Danger Group 1992) [13].

The FDRS that was developed for South-east Asia (Indonesia and Malaysia) have structure similar to the FWI System. However, agency-specific differences in forest and land fire management policy may result in minor differences in FDRS applications. For example, FDRS decision aids which detail fire management actions such as fire prevention, detection, or suppression mobilization activities, may occur at different fire danger levels in Indonesia and Malaysia because of local fire issues and fire management guidelines. The main tasks of the Southeast Asia Fire Danger Rating System Project were to conduct the initial calibration of the FWI System to regional conditions, to facilitate development of FDRS-based fire management decision-aids to address local fire problems, to strengthen technical development, coordination, management and integration of fire systems in the region [4].

However, FDRS parameters of Malaysia are forecasted based on interpolated data gathered from a pool of national weather stations and does not necessarily reflect the actual local climate in the forest. In addition, the FDRS parameters are extracted based on atmospheric conditions and does not include ground data such as Ground Water Level (GWL) which are one of crucial factors for peatland forest fire planning and management. Currently at least half of all peatland areas are drained with varying intensity [7,8]. As a result, peatland distribution and peatland management have played major roles in influencing insular Southeast Asian fire regions. The most important approach in this regard is ground water level management. If water levels within fire prone peatlands are maintained at a high level through the dry season, the risk of significant fires can be reduced. This can be done through blocking of any drainage ditches in peatland forest or abandoned agricultural land and improving the water management in any existing plantation and agricultural land through installation and operation of water control structures. Where possible water levels should be maintained in natural peatland forest areas at no more than 20 cm below the surface and in agriculture and plantation areas, no more than 40 cm below the surface [1]. In high fire risk areas options to enhance the water level through tapping artesian water through tube wells or pumping water from rivers or lakes can be considered [1,13]. Meanwhile, for the already deforested degraded undeveloped peatland areas there are two theoretical possible opposing management directions to reduce the risk of fire episodes; (1) peat swamp rehabilitation including blocking of the drainage canals [6] and (2) conversion to managed agricultural areas with presumably higher incentive for fire control. In managed areas (e.g., industrial plantations), effectiveness of fire prevention measures taken and capabilities to retain the highest possible water table levels tolerated by the plantation species are seen as key features in fire risk reduction [9].

However, due to lack of manpower to obtain in-situ samples for analysis and monitoring purpose, IoT is required to better prepare for immediate fire. As a *de facto* technology for IoT networks, LoRa is a low-power wide area network protocol developed based on spread spectrum modulation techniques. Our previous work has shown that transmission design parameters can be configured for mangrove monitoring [5]. LoRa transmission parameters such as spreading factor, coding rate, data rate and packet size can be optimised in order to send multimedia traffic of the area of interest. This paper proposes an IoT-based system for peatland monitoring with the aim to provide efficient communication between land and fire managers with real-time monitoring tower equipped with sensors and wireless network. IoT system involves data collection, transmission storage, operating systems, and applications [2].

This paper reports on IoT-based peatland system deployed in Raja Musa Forest Reserve (RMFR), Kuala Selangor, Malaysia. The data reported in this study was analysed for middle January 2020 until end of March 2020 to highlight initial findings of this project implementation. The project is part of an international initiative where sensory data of peatland sites at three ASEAN countries (Malaysia, Indonesia and Brunei) are collected and connected to a central cloud hosted by NICT Japan. This type of IoT-based peatland systems could support integrated management plan for North Selangor Peat Swamp Forest for 2014 to 2023 [10].

2 IoT-Based Peatland Forest Management and Monitoring System

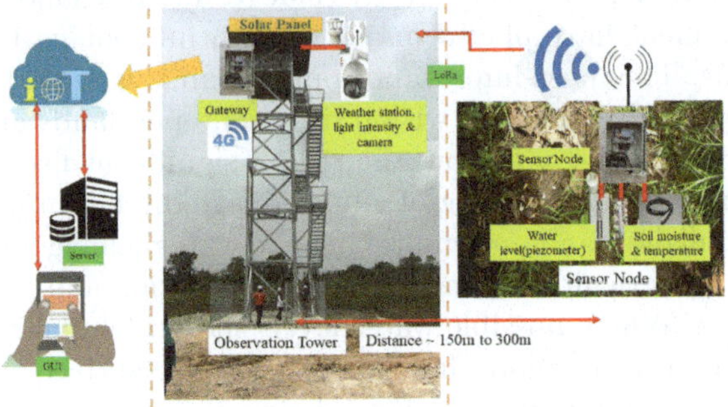

Fig. 1. IoT deployment setup in Raja Musa Forest Reserve (RMFR), Kuala Selangor, Malaysia.

The IoT-based peatland forest management and monitoring system was carried out at Raja Musa Forest Reserve (RMFR), Kuala Selangor, Malaysia. Several forest burning incidents raged the forest area such as during the 98-*El-Niño* in 1998 and several other incidents in 2000, 2005 and 2017 [8]. The deployment setup is shown in Fig. 1.

In Fig. 1, the IoT system collects peatland data from two ground sensor nodes measuring soil temperature, soil humidity and ground water level (GWL). Both ground sensor nodes are connected to a gateway via LoRa access technology. Then, the peatland data is transmitted to the cloud using 4G mobile networks. The cloud server is located at MIMOS Malaysia for the initial implementation. Data at this point is also accessible to the stakeholders from the local forest management and community. NICT server consolidates all data from the participating ASEAN countries (i.e. data from peatlands in RMFR of Malaysia, Badas of Brunei and Jambi of Indonesia). Since LoRa transmission range depends on its propagation condition and in this case the signal is heavily distorted by the peatland bushes and trees (approximately of 2–10 m height and of various density), the gateway itself is placed on a tower approximately 25 m of height above ground level. A weather station is also installed near the gateway which provides local atmospheric sensory data including that information necessary to derive FDRS parameters (i.e. wind speed, rain precipitation, ambient temperature and ambient humidity). A CCTV camera is installed at the tower to provide visual surveillance of the equipment and the area. All equipment at the gateway are powered using solar panel, whereas the ground sensor nodes are battery operated. In order to monitor the signal strength and the batteries health, the ground sensor nodes also transmit information about their current Signal to Noise Ratio (SNR), Received Signal Strength Indicator (RSSI) and battery status. The chosen locations are amongst the hot spots and are prone to forest fire in RMFR which requires constant monitoring. The observation gateway has a latitude and longitude of (3.46595, 101.441388). The first ground sensor node is located 180 m away from the gateway. Another sensor node which is 200 m towards south represents reading at a lower gradient. There is a man-made water canal parallel to the access road which is near to sensor node 1 (SN1). On the other hand, sensor node 2 (SN2) is located deeper in the peat forest, further from the access road and the canal. Figure 2 illustrates the location of both ground sensor nodes relative to the observation tower.

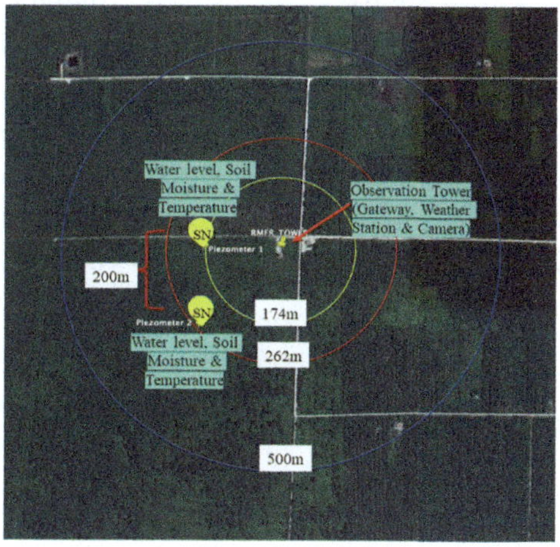

Fig. 2. Relative locations of the ground sensor nodes (SN1 and SN2) from the observation tower.

The parameters measured at weather station and deployed sensors are listed in Table 1 and 2, respectively. Table 3 shows the fire risk code based on the water depth range.

3 Analysis of RMFR Ground Water Level

The ground water level or water table is measured using a piezometer. A borehole is prepared first for installation of the piezometer sensor. Piezometer sensor measures the amount of water pressure and converted into water level. A UPVC pipe install in the borehole and terminated down to the mineral soil layer with approximately around 5.6 m as depicted in Fig. 3. This to ensure that the measured water level is the actual water level within the peat layer. Surrounding the pipe, small holes are made to ensure water at the peat layer could penetrate and fill in within the UPVC pipe. The perimeter surrounding the boreholes are guarded from wild animals using a 1 m height fence. In this study, we focus mainly on the mean water level measured and the mean water level difference for the peatland monitoring.

Table 1. Measurement parameters at weather stations.

Parameters	Unit	Range
Date and time	dd/mm/yy, hh:mm:ss	–
Accumulated precipitation	mm of liquid water	0–500 mm
Accumulated solar radiation	KJoule	Daily Solar Radiation KJ
Air pressure	hPa	150–1100 hPa
Wind speed	m/s	0–60 m/s
Humidity	%RH	0–100%
PM_{10} concentration	$\mu g/m^3$	0–1000 $\mu g/m^3$
$PM_{1.0}$ concentration	$\mu g/m^3$	0–1000 $\mu g/m^3$
$PM_{2.5}$ concentration	$\mu g/m^3$	0–1000 $\mu g/m^3$
Precipitation intensity	mm/hr	0–100mm/hr
Radiation illuminance	Lux	0–20 K Lux
Solar radiation power	W/m^2	0–2000 W/m^2
Temperature	° C	−40° C– +80° C
UV radiation	W/m^2	0–2000 W/m^2
Wind direction	°	0°–359°

Table 2. Measurement parameters at ground sensor node 1 and node 2.

Parameters	Unit	Range
Date and time	dd/mm/yy, hh:mm:ss	–
Battery	Volt	0–3.7 V
Depth	mm	0–400 mm
Soil moisture	%	0–100%
LoRa Signal to Noise Ratio (SNR)	dB	0dB–10dB
Latitude	°	0°–180°
Longitude	°	0°–180°
RSSI	dBm	0dBm–46dBm
Soil temperature	° C	−40° C– +180° C

Table 3. Fire risk codes.

Water table depth range (mm)	Colour code	Fire risk
500 to 0	Blue	Low
−500 to 0	Green	Medium
−500 to −700	Yellow	High
−700 to −1000	Red	Extreme

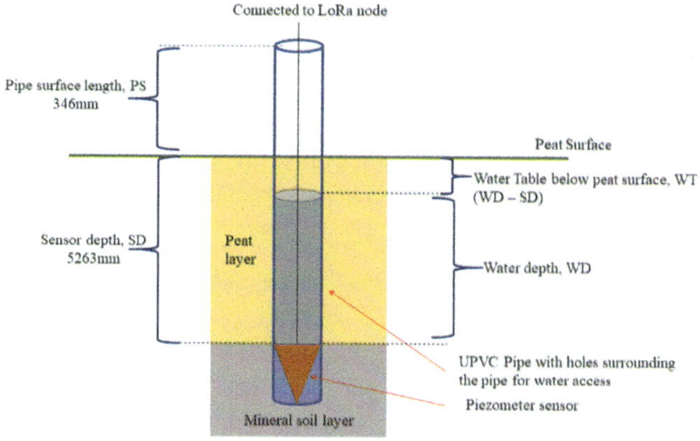

Fig. 3. Illustration of piezometer sensor and UPVC pipe installation to measure ground water level.

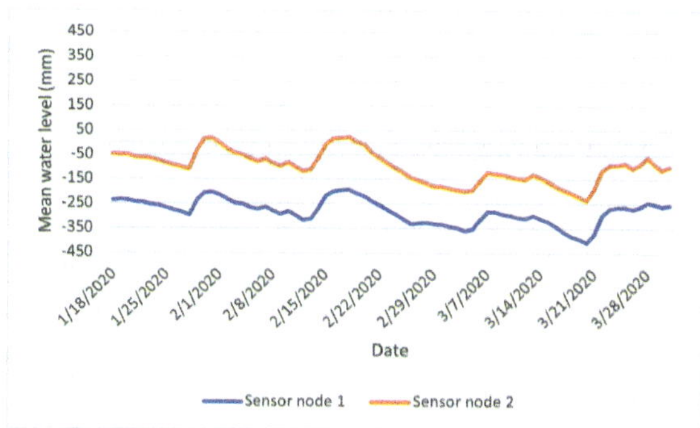

Fig. 4. Mean water table recorded from January to March 2020.

Figure 4 illustrates preliminary results for the IoT system deployment for both piezometers of the sensor nodes. The data has been collected between the periods of mid-January 2020 to end of March 2020. The two trend lines reflect similar curves with a high Pearson correlation value of 0.97. This implies that the readings of both sensors are coherent with each other. However, there is a consistent difference margin of approximately around 220 mm between the two lines where readings from sensor node 1 is consistently lower than its counterpart. A possible explanation to this is that the location of sensor node 1 which is near to the canal, hence water is consistently being drawn out from its location. Another possible explanation is variance in the elevation of the land. The preliminary results of Fig. 4 also indicate that the water level falls within Medium (i.e. Green) Fire Risk Codes (i.e. between 0 mm to −500 mm). The period within the month February to April is mostly dry season in Malaysia which explains the lower mean water level. According to a study of [11] dry period for the Peninsula Malaysia usually occurs during the Southwest Monsoon (SWM). In general,

Peninsular Malaysia climate is influenced by the Southwest monsoon, from May to August, and the Northwest Monsoon (NWM), from November until February [11]. The weather station sensory data records mean temperature and relative humidity of 27.6 °C and 81.8% respectively within this period with median rainfall of only 0.6 mm. More observations need to be carried out to study the trends of peatland data throughout various seasons.

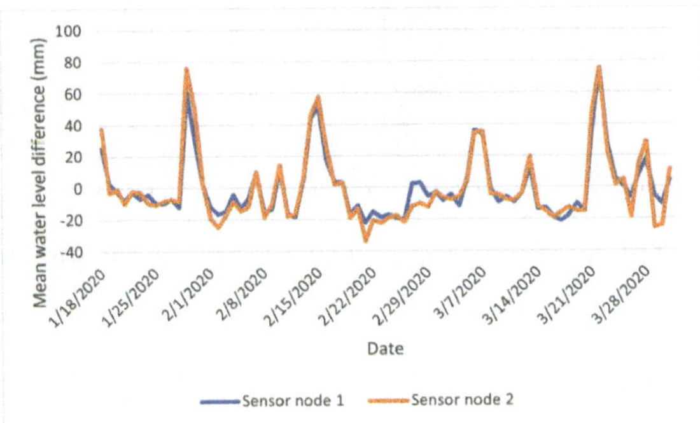

Fig. 5. Mean water level difference recorded from January to March 2020.

Figure 5 shows the mean water level difference for the ground sensor nodes deployed. The mean water level difference is calculated based on the difference between the measured mean water level on the day and that of the previous day. This parameter can provide insight on the trend of water level of the peatland that can be affected by different/multiple reasons. Upon the analysis, the water level difference of sensor node 1 and 2 shows high level of correlation with a Pearson correlation coefficient of 0.96. Due to the relatively close location of the deployed sensor nodes, 200 m apart as shown in Fig. 2, the received rainfall is almost identical. Therefore, the water level difference for the same date are almost indistinguishable.

To employ a more efficient peatland monitoring, it is vital to study the relationship between the data collected by the sensor nodes and the data collected by the weather station. Understanding of the correlation of these data allows for detailed insight on the behaviour of local climate and potential prediction of certain parameters using the others or vice versa. Table 4 tabulates the result of correlation analysis after comparing the measured parameters between weather station to mean water level and mean water level difference, respectively. Pearson correlation coefficient is a correlation coefficient commonly used to study the linear correlation between two variables. Pearson correlation coefficient ranges from -1 to $+1$, where coefficient closer to -1 or $+1$ means stronger strength of association, whilst coefficient closer to zero means absence of linear relationship between the variables.

From Table 4, it is noticed that concentration of pollutants (PM_{10}, PM_1, $PM_{2.5}$) shows a negative correlation with the mean water level, exceeding -0.5

Table 4. Pearson correlation analysis

	Mean water level		Mean water level difference	
Weather station	Sensor node 1	Sensor node 2	Sensor node 1	Sensor node 2
Air Pressure	0.3414	0.4403	0.1003	0.1335
Ambient Humidity	0.1250	0.1026	0.6572	0.6754
PM_{10} Concentration	−0.5173	−0.5347	−0.0876	−0.0770
PM_1 Concentration	−0.5012	−0.5208	−0.0931	−0.0824
$PM_{2.5}$ Concentration	−0.5135	−0.5324	−0.0873	−0.0762
Precipitation Intensity	0.1155	0.1202	0.6576	0.6702
Radiation Illuminance	−0.0911	−0.1199	−0.5120	−0.5112
Solar Radiation	−0.0911	−0.1199	−0.5120	−0.5113
Ambient Temperature	−0.2390	−0.2741	−0.6230	−0.6265
UV Radiation	−0.1155	−0.1415	−0.5297	−0.5327
Wind Speed	0.0526	0.0301	−0.0162	−0.0903
Accumulated Rainfall	0.3299	0.3495	0.8305	0.8724

for both sensor nodes. However, the most interesting observation comes from the comparison of weather station with the mean water level difference. The mean water level difference for sensor node 1 and 2 have very strong positive correlation with the accumulated rainfall measured by the weather station, which are 0.8305 and 0.8724 respectively. This phenomena is due to the fact that, higher amount of rainfall can definitely increase the amount of water in the soil, hence, the increases of water level compared to the previous day.

4 Conclusion

IoT system can be used to monitor peatland especially in forest management. The paper has presented a deployed IoT system at Raja Musa Forest Reserve (RMFR), Selangor, Malaysia. The sensors are installed and connected to the cloud server via LoRaWAN and cellular networks. The peatland data from January to March 2020 are analysed to highlight preliminary findings for this project implementation. This pilot deployment opens up more opportunities for the local authorities and communities to better manage peatland forests by both in-situ and IoT-based monitoring. This will in turn reduce the potential of forest fires which could lead to transboundary haze in ASEAN countries.

Acknowledgement. The authors would like to acknowledge NICT Japan and ASEAN-IVO for funding this project NAPC (Networked ASEAN Peatland Forests Communities), Selangor State Forestry Department (JPNS) for permission to deploy the IoT system, MetMalaysia, LORANET Technologies and Global Environment Centre (GEC) for valuable feedback and validation of peatland data.

References

1. Agency, N.E.: ASEAN guidelines on peatland fire management. Haze Action Online (2015)
2. Ayaz, M., Ammad-Uddin, M., Sharif, Z., Mansour, A., Aggoune, E.H.M.: Internet-of-things (IoT)-based smart agriculture: toward making the fields talk. IEEE Access **7**, 129,551–129,583 (2019)
3. Barber, C.V., Schweithelm, J.: Trial by Fire. World Resources Institute, Washington, D.C. (2000)
4. De Groot, W.J., Field, R.D., Brady, M.A., Roswintiarti, O., Mohamad, M.: Development of the Indonesian and Malaysian fire danger rating systems. Mitig. Adapt. Strat. Glob. Change **12**(1), 165 (2007)
5. Jebril, A.H., Sali, A., Ismail, A., Rasid, M.F.A.: Overcoming limitations of LoRa physical layer in image transmission. Sensors **18**(10), 3257 (2018)
6. Miettinen, J., Hooijer, A., Vernimmen, R., Liew, S.C., Page, S.E., et al.: From carbon sink to carbon source: extensive peat oxidation in insular Southeast Asia since 1990. Environ. Res. Lett. **12**(2), 024014 (2017)
7. Miettinen, J., Shi, C., Liew, S.C.: Land cover distribution in the peatlands of Peninsular Malaysia, Sumatra and Borneo in 2015 with changes since 1990. Glob. Ecol. Conserv. **6**, 67–78 (2016)
8. Nuruddin, A.A., Leng, H.M., Basaruddin, F.: Peat moisture and water level relationship in a tropical peat swamp forest. J. Appl. Sci. **6**(11), 2517–2519 (2006)
9. Page, S.E., Hooijer, A.: In the line of fire: the peatlands of Southeast Asia. Philoso. Trans. Royal Soc. B: Biol. Sci. **371**(1696), 20150176 (2016)
10. Parish, F., Dahalan, M., Rahim, H.: Integrated management plan for North Selangor peat swamp forest 2014–2023 for Selangor State Forestry Department. Draft Revision **2** (2014)
11. Suhaila, J., Deni, S.M., Zin, W.Z.W., Jemain, A.A.: Trends in peninsular Malaysia rainfall data during the Southwest monsoon and Northeast monsoon seasons: 1975–2004. Sains Malaysiana **39**(4), 533–542 (2010)
12. Vitolo, C., Di Giuseppe, F., Krzeminski, B., San-Miguel-Ayanz, J.: A 1980–2018 global fire danger re-analysis dataset for the Canadian fire weather indices. Sci. Data **6**, 190032 (2019)
13. Wösten, J., Clymans, E., Page, S., Rieley, J., Limin, S.: Peat-water interrelationships in a tropical peatland ecosystem in Southeast Asia. Catena **73**(2), 212–224 (2008)
14. Yule, C.M.: Loss of biodiversity and ecosystem functioning in Indo-Malayan peat swamp forests. Biodivers. Conserv. **19**(2), 393–409 (2010)

Author Index

W. S. H. Suhaili et al. (Eds.): CIIS 2020, AISC 1321, pp. 243–244, 2021.
https://doi.org/10.1007/978-3-030-68133-3

The manufacturer's authorised representative in the EU is Springer
Nature Customer Service Centre GmbH, Europaplatz 3, 69115 Heidelberg,
Germany. If you have any concerns regarding our products, please
contact ProductSafety@springernature.com

Printed and bound by CPI Group (UK) Ltd, Croydon, CR0 4YY
28/09/2026
02224073-0007